Lecture Notes in Mathematics　1828

Editors:
J.-M. Morel, Cachan
F. Takens, Groningen
B. Teissier, Paris

T0218375

Springer
Berlin
Heidelberg
New York
Hong Kong
London
Milan
Paris
Tokyo

David J. Green

Gröbner Bases and the Computation of Group Cohomology

 Springer

Author

David J. Green

FB C - Mathematik und Naturwissenschaften
Bergische Universität Wuppertal
Gaussstr. 20
42097 Wuppertal
Germany
e-mail: green@math.uni-wuppertal.de

Cataloging-in-Publication Data applied for
Bibliographic information published by Die Deutsche Bibliothek

Die Deutsche Bibliothek lists this publication in the Deutsche Nationalbibliografie;
detailed bibliographic data is available in the Internet at http://dnb.ddb.de

Mathematics Subject Classification (2000): 20J06, 16S15, 16E05, 16Z05, 20C05, 20D15

ISSN 0075-8434
ISBN 3-540-20339-7 Springer-Verlag Berlin Heidelberg New York

Springer-Verlag is a part of SpringerScience+Business Media

springeronline.com

© Springer-Verlag Berlin Heidelberg 2003
Printed in Germany

The use of general descriptive names, registered names, trademarks, etc. in this publication does not imply,
even in the absence of a specific statement, that such names are exempt from the relevant protective laws
and regulations and therefore free for general use.

Typesetting: Camera-ready TEX output by the authors
SPIN: 10964583   41/3142/du - 543210 - Printed on acid-free paper

For Birgit, Thomas and Anne

Preface

The motivation for this book is the desire to perform complete calculations of cohomology rings in the area called the cohomology of finite groups. The theories presented here belong to that part of computational algebra known as noncommutative Gröbner bases. It happens that existing Gröbner basis methods perform particularly poorly in the conditions imposed by cohomology calculations. So new types of Gröbner bases had to be developed, informed by practical computability considerations.

Thanks to the work of J. F. Carlson, one can compute the cohomology ring of a p-group from a sufficiently large initial segment of the minimal projective resolution. The first new Gröbner basis theory – for modules over the modular group algebra – was developed to construct the minimal resolution as efficiently as possible. In all probability it applies equally well to finite dimensional basic algebras.

Carlson's method also needs the ability to manipulate the relations in the cohomology ring. As such rings are graded commutative rather than strictly commutative, it was necessary to devise a theory of Gröbner bases for graded commutative rings. There is more than one way to do this. The Gröbner bases presented here were designed to resemble the classical commutative case as closely as possible. Strictly speaking, they are Gröbner bases for right ideals in a more general type of algebra which is here called a Θ-algebra.

Many cohomology computations have been performed using these methods. In particular, the essential conjecture of Mui and Marx was shown to be false. The counterexample is the Sylow 2-subgroup of $U_3(4)$, a group of order 64.

Acknowledgements

I am very grateful to the following people.

Jon F. Carlson aroused my interest in the computer calculation of group cohomology. All the big computations described here were performed on his machine toui.

Klaus Lux and Peter Dräxler drew my attention to the potential of noncommutative Gröbner basis methods, and in particular to the work of Ed Green and his collaborators.

Prof. G. Michler (who suggested computational cohomology to me as long ago as 1991) allowed me the use of the computers at the Institute for Experimental Mathematics in Essen during the long developmental phase.

This book is my Wuppertal *Habilitationsschrift*, with minor updates. But for the exertions of Erich Ossa and Björn Schuster, the number of grammatical errors in the German original would have been considerably higher.

Wuppertal, August 2003   *David J. Green*

Table of Contents

List of symbols

Introduction

The motivation for this book is the desire to perform complete calculations of cohomology rings in the cohomology theory of finite groups. The theories presented here belong to that area of computational algebra known as non-commutative Gröbner bases. It happens that existing Gröbner basis methods perform particularly poorly here, and so new types of Gröbner bases had to be developed, informed by practical computability considerations.

Let p be a prime number, G a finite p-group, and k a field of characteristic p. We want to determine the mod-p cohomology ring $H^*(G, k)$ for as many such G as possible. One reason one should want to do this is in order to develop and test conjectures about the structure of cohomology rings. For example, in §7.1.3 we shall refute the conjecture of Marx and Mui on the nilpotency degree of the ideal of essential classes.

One reason for restricting our attention to p-groups is that there are many methods for determining the mod-p cohomology ring of a finite group from the cohomology of its p-local subgroups. The first was the stable element method of [24, XII §10], whereas more recent results have a stronger geometric flavour [58, 29, 30, 44]. Another reason is that p-groups are well suited to the minimal resolutions approach described below. This is because there is only one simple module – the trivial module – and all projective modules are free.

Broadly speaking there are three available approaches for calculating the cohomology rings of all p-groups of a given size.

1. Calculation by hand
   The method of calculation has to be tailored to the individual group under consideration. Straightforward to write up the calculation so that it can (in theory) be checked by the mathematical public, but surreptitious errors can (and do) creep in.
2. Computer calculation using the Eilenberg–Moore spectral sequence
   Notably used by Rusin in [57] to determine the cohomology rings of all 51 groups of order 32. Seems not to be suitable for the case $p > 2$.
3. Computer calculation using sufficiently many terms in the minimal projective resolution of the kG-module k.
   Recent work of Carlson [18] means that one can now detect when enough terms have been constructed. More recently still Carlson determined the cohomology rings of all groups of order 64 by this method [19, 20].

Here we shall adopt the minimal resolutions approach. Following Carlson [21, 18] this consists of three stages:

1. One constructs the first N terms in the minimal resolution. Choosing N requires some care.
2. By lifting cocycles to chain maps one computes products and so determines a partial presentation of the cohomology ring, consisting of generators and relations in degree at most N.
3. If this presentation passes a series of tests which are described in [18] then there are no new generators or relations in higher degrees and the partial presentation of the cohomology ring is complete. If it fails any of the tests then try again with a larger value of N.

We shall refer to this series of tests as Carlson's Completeness Criterion.

This book is concerned with the question of how best to put this minimal resolutions approach into practice. In particular we are interested in efficient methods for constructing minimal resolutions and for working with presentations of (graded commutative) cohomology rings.

To construct the minimal resolution one needs a highly efficient method for computing minimal generators for the kernel of a map between free modules over the modular group algebra kG. A beautifully simple linear algebra approach is described by Carlson et al. in [21]. This method is easy to program and is unbeatable for small 2-groups. However as one considers larger and larger groups, this approach becomes less and less efficient. The attraction of (noncommutative) Gröbner basis methods is that they allow the computer to work linearly over the group ring kG, rather than linearly over the ground field k as in the elementary linear algebra approach. This does of course mean that the programs take much more work to write, but the space savings for large groups are considerable (see e.g. Example 0.2 on page 5). The methods developed here are tailor-made for working with modules over p-groups. Note that general-purpose noncommutative Gröbner basis methods already exist for constructing minimal resolutions, even over basic algebras [34, 43]. Moreover there do exist other approaches to the construction of (near-)minimal projective resolutions over modular group algebras of p-groups [37, 59].

The first part of this book introduces a new kind of Gröbner bases for modules over p-groups and establishes a new, two-speed version of Bergman's Diamond Lemma [12]. A new way to construct minimal projective resolutions is presented which uses these Gröbner bases. A second new kind of Gröbner bases is introduced in the second part: Gröbner bases for graded commutative algebras. These make it possible to perform computer calculations in group cohomology rings at odd primes, for there the cohomology rings are graded commutative rather than strictly commutative. I have developed a package called Diag which uses these two kinds of Gröbner bases to compute cohomology rings. Results obtained with this package are described in the third part of the book. In particular, the cohomology rings of all 15 groups of order 81 were computed, and the cohomology ring of the Sylow 2-subgroup of $U_3(4)$ was found to contain two essential classes with nonzero product.

Gröbner bases

There are Gröbner bases for many different kinds of algebras and modules. Essential components of any Gröbner basis theory are

- A canonical form for elements of the algebra or module, which can be calculated using a so-called Gröbner basis.
- The Buchberger Algorithm, an algorithm which constructs a Gröbner basis.

Gröbner bases are important because of their applications to computer calculations, for a Gröbner basis allows the computer to work linearly over an algebra (not just over a field).

Gröbner bases are best known in commutative algebra: textbooks here include Adams–Loustaunau [1] and Eisenbud [31]. But there are also Gröbner bases in noncommutative algebra. Here we shall introduce Gröbner bases for modules over the modular group algebra of a p-group and for graded-commutative algebras.

Gröbner bases for modules over p-groups

In its original form the Diamond Lemma is a result about associative algebras. Let $\Lambda = k\langle a_1, \ldots, a_r\rangle/(B_1 - h_1, \ldots, B_s - h_s)$ be a presentation of such an algebra, where each B_i is a monomial. The normal form of an element $f \in \Lambda$ is obtained by repeatedly substituting h_i for B_i in a monomial in the support of f until this is no longer possible. It is assumed that each f is reduction-finite, i.e., that every such sequence of "reductions" is finite. The Diamond Lemma then gives conditions under which the normal form of f is independent of the sequence of reductions employed. If so, the image of f in Λ is zero if and only if the normal form of f is zero. The generators $(B_i - h_i)_{i \leq s}$ of the relations ideal are then said to form a Gröbner basis for Λ.

Bergman's original paper [12] also sketches the extension of the Diamond Lemma to Λ-modules. Let M be a submodule of the free right Λ-module F. Both the relations in Λ and the generators of M give rise to reductions. However if Λ is the modular group algebra kG of a p-group, or another algebra of small dimension in which it is easy to calculate, then it is tempting to assign the relations in Λ a higher priority than the generators of M. This leads to a two-speed reduction system, and there are many examples where some elements are only reduction-finite for the two-speed system.

Example 0.1. A presentation of $kG = \mathbb{F}_3 C_3$ ist $k[a]/(a^3)$. Let F be the free kG-module on one generator e and let M be the submodule of F generated by $ea - ea^2$. Then $ea \in M$, because $ea = (ea - ea^2)(1 + a) + ea^3$. So in a good reduction system, ea should be reduction-finite with normal form 0.

The two reductions are: substitite 0 for a^3, and ea^2 for ea. If both reductions have the same priority then ea is not reduction-finite, for repeatedly substituting ea^2 for ea leads to the sequence

$$ea \mapsto ea^2 \mapsto ea^3 \mapsto ea^4 \mapsto \cdots .$$

However if the reduction $a^3 \mapsto 0$ is given the higher priority then $ea^3 \mapsto 0$ is the only allowable reduction of ea^3, and so $ea \mapsto ea^2 \mapsto ea^3 \mapsto 0$ is the only allowable sequence of reductions for ea. So ea is reduction-finite with normal form 0.

The Diamond Lemma for two-speed reduction systems is proved as Theorem 2.12. It is assumed that a *negative* word ordering is being used, so that 1 is the largest and not the smallest word. A negative word ordering was tacitly used in the above example, for the module generator was interpreted as the reduction $ea \mapsto ea^2$ and not as the reduction $ea^2 \mapsto ea$. Negative orderings are unusual for a Gröbner basis theory, as such orderings are not well-orderings. Bergman's Diamond Lemma presupposes a well-ordering too. However the two-speed reduction and the fact that the underlying algebra is finite-dimensional do in fact allow us to prove the Diamond Lemma for a negative ordering. And negative orderings have two significant advantages: they lead to simpler presentations of the group algebra, and they are essential for the new method for constructing minimal resolutions.

As one would expect there is a Buchberger Algorithm (Algorithm 2.20) that constructs a Gröbner basis for a given submodule of a free Λ-module F. Here the two-speed reduction has a significant advantage for computer calculation: one works throughout with a fixed k-basis of F, namely the basis $E_{\mathbf{X}}(F)$ of Definition 1.11.

Gröbner bases and minimal resolutions

As G is a p-group and k has characteristic p, there is only one projective indecomposable kG-module: namely kG itself. This means that minimal resolution of the trivial module can be constructed by iterating the following key task:

> Let ϕ be a kG-linear map between free kG-modules. Determine **minimal generators** for the **kernel** of ϕ.

Gröbner basis methods already exist for this key task. For example, C. Feustel, E. Green and their coworkers have developed a Gröbner basis theory for basic algebras[1] and a method for constructing minimal resolutions using these Gröbner bases [33, 34].

---

[1] Algebras whose simple modules are all one-dimensional, such as modular group algebras of p-groups.

But general Gröbner basis methods have to be applied very carefully if they are to work for modules over p-groups, for Gröbner bases were primarily developed to be able to work in infinite-dimensional algebras. And although the methods for basic algebras were of course developed for the finite-dimensional case, it turns out that group algebras of p-groups are pretty pathological basic algebras[2].

Here we shall use the two-speed Diamond Lemma to derive a new Gröbner basis method for constructing minimal resolutions. As one would expect, one can use elimination (a variant of the Buchberger Algorithm) to determine a Gröbner basis for the kernel of a module homorphism. However this will usually not be a minimal generating set of the kernel. So for the second part of the key task we shall derive a second variant of the Buchberger Algorithm which computes minimal generators of a given submodule M of a free module F. Here it is essential that the underlying word ordering is negative, for then it is particularly easy to determine the radical of a module. This second Buchberger variant calls for another new version of the Diamond Lemma (Theorem 3.12), which even allows for a three-speed reduction. For the reductions coming from generators of M have to be split into two classes, with the reductions coming from elements of the radical receiving the higher priority.

Example 0.2. Let G be a Sylow 2-subgroup of the sporadic Mathieu group M_{24}. This G has order 2^{10} and 2-rank 6. The 7th and 8th terms in the minimal resolution are free of rank 222 and 336 respectively. To calculate the 9th term one has to determine the kernel of the 8th differential d_8. One possibility would be to construct the matrix of d_8 over k and take its null space. This matrix requires

$$222 \times 336 \times 2^{20} \text{ entries} = 9,11 \text{ GB.}$$

The matrix is neither sparse nor are any other structural properties known that could allow us to compress it significantly.

The new Gröbner basis method was used to construct the minimal resolution out to the 9th term. The 9th term is free of rank 485. The Gröbner basis used in the elimination requires 130 MB and the Gröbner basis for minimal generators requires 124 MB. The 9th differential d_9 requires 20 MB. Computing d_9 from d_8 took 6 days 18 hours.

Growth rate of minimal resolutions

Let G be a p-group of p-rank r. That is, the largest elementary abelian subgroups of G have order p^r. Each term in the minimal projective resolution of k over kG is a free module, and one knows (see [6, §5.3]) that the rank of this

---

[2] There is only one simple module, and there are very long paths in the path algebra whose images in the basic algebra are nonzero.

free module grows as a polynomial of degree $r-1$. Turning now to the problem of constructing enough terms in the minimal resolution (enough in the sense of Carlson's Completeness Criterion), one sees that the complexity increases very quickly when the size of the groups under investigation increases. For the dimension of the group algebra (the unit in terms of which the above growth rate is measured) increases by a factor of p, and typically both r and the number of terms that counts as enough increases. For odd primes the situation is even worse than for 2-groups: the dimension of the group algebra increases much faster, and many more terms in the minimal resolution are required.

Example 0.3. Consider the extraspecial p-group $G = p_+^{1+2}$ of order p^3. For $p = 2$ this is the dihedral group D_8, whereas for p odd it has exponent p. For $p = 2$, the presentation of the cohomology ring attains completeness in degree 2, and Carlson's criterion detects this in degree 4. For $p = 3$, completeness is attained in degree 6 and detected in degree 8. For $p \geq 5$, there is a relation in degree $4p - 3$ (see [46]).

Gröbner bases for graded-commutative algebras

A Gröbner basis theory for graded-commutative algebras is set up in Chap. 4, since it appears this has not been done before. The computer needs this kind of Gröbner basis to be able to work in graded-commutative algebras, such as cohomology rings in odd characteristic.

Here there is more than one possible theory of Gröbner bases. If y is an odd-dimensional element of a graded-commutative algebra A then the relation $y^2 = 0$ holds automatically in A. The theory developed here was chosen to treat these "structural" relations as far as possible like ordinary relations, and to resemble the classical commutative case as closely as possible. Such a theory is indeed possible, in spite of some difficulties (see cautionary Example 4.33).

Cohomology rings of p-groups

Cohomology of finite groups is an area that lies at the intersection of several branches of mathematics. Here algebraic topology, group theory and modular representation theory all meet. Modern texts include Evens' book [32] and Benson's book [6].

Cohomology rings of p-groups play a significant role as they are the "atoms" of cohomology theory[3]. One way to see this is to consider modular representation theory. The representation theory of a group is determined by the simple modules together with the various ways of putting simple modules

---

[3] In the sense that they cannot be simplified any further

together to form an indecomposable module. The cohomology of the group describes – admittedly in coded form – the ways to put modules together. Now for p-groups there is only one simple module and this is trivial. Hence the representation theory just concerns the ways to stick trivial modules together: that is, the cohomology.

Methods for calculating cohomology

Several elegant local methods for determining the cohomology ring have been developed over the last fifteen years [58, 29, 30, 44]. But it is in the nature of a local method that it says nothing about the case of a p-group and rather assumes that the cohomology rings of the p-local subgroups are known. For a long time spectral sequences were the only way to calculate the cohomology ring of a p-group. But spectral sequences are not suitable for computing the cohomology rings of all p-groups of a given order, Rusin's notable success [57] with the groups of order 32 notwithstanding. For even to compute the cohomology rings of all groups of order 5^3 one has to use methods that are tailored to the particular group under consideration and are sometimes indirect (see Leary's paper [46]). For many groups of order p^4 or p^5 the spectral sequences are still unsolved or only partially solved.

The article [18] by J. F. Carlson is a turning point, for it describes a method for the systematic computer calculation of group cohomology. To be more precise, an old method is made computationally feasible for the first time. Carlson shows how to compute the cohomology ring of an arbitrary p-group given sufficiently many terms in the minimal projective resolution of the trivial module.

On the one hand this method may be called old, for the algebraic definition of the cohomology ring is based on a projective resolution of the trivial module. However only the cohomology rings of the cyclic groups were computed in this way, for the minimal resolution of such a group is periodic and can therefore be described completely in finite time. But for an arbitrary p-group one can only hope to construct finitely many terms in the minimal resolution. Admittedly cohomology rings are finitely presented, and one may compute the product structure out to degree N given the minimal resolution out to the Nth term. But previously there was no method to decide whether all generators and relations had been found, or whether there were new generators and/or relations in higher degrees. No (useable) degree bounds are known.

What is new in Carlson's method is a criterion that allows us to conclude that the generators and relations out to degree N do give us a complete presentation of the cohomology ring. Recently Carlson finished computing the cohomology rings of all groups of order 64 using this method. It is not yet proven that this method works for all p-groups. More precisely, there could conceivably be groups G for which the Carlson's Completeness Criterion is never satisfied, no matter how large an N one takes. But the criterion is a sufficient condition: if it is satisfied, then the presentation has been proved to be complete.

Summing up, Carlson's Completeness Criterion opens up the minimal resolution as a way to compute the cohomology ring.

Implementation

The method for constructing minimal resolutions was implemented in the C programming language, using the library of the package "C MeatAxe" [56] to work with vectors over $\mathbb{F}_p$. Moreover a Gröbner basis implementation of Carlson's method for computing cohomology rings was developed, again in C. The computer algebra system GAP [36] was used to assemble the necessary facts about the group and its subgroup structure. The package is called Diag. I am going to create a revised and documented version for public release, but in the meantime interested persons are welcome to contact me at email address green@math.uni-wuppertal.de for a copy of the current (unpolished) version.

Experimental results

To date the applications have been in two areas:

— Computing cohomology rings of small p-groups.
— Constructing as many terms as possible in the minimal resolution for larger p-groups.

All big computations were performed on Jon F. Carlson's computer toui, a Sun ULTRA 60 Elite 3D.

Additionally the exact period of one periodic module was calculated to give a taste of another area where the package could be used.

Small groups

To date the main results are:

— The cohomology rings of all 15 groups of order $81 = 3^4$.
— The cohomology rings of 14 of the 15 groups of order $625 = 5^4$.
— The cohomology ring of the Sylow 2-subgroup of $U_3(4)$, a group of order $64 = 2^6$. For the first time two essential classes with nonzero product were found in a group cohomology ring.
— The cohomology ring of a certain group of order $243 = 3^5$. This ring has Krull dimension 3 and depth 1. This is the first completely calculated case at an odd prime where the dimension exceeds the depth by at least two.

These cohomology rings and more are available on the World Wide Web at the following address:

```
http://www.math.uni-wuppertal.de/~green/Coho/index.html
```

Minimal resolutions

Three groups of order 2^9 or 2^{10} were studied which arise as Sylow 2-subgroups of sporadic finite simple groups.

- The Sylow 2-subgroup of the Higman–Sims group HS has order 2^9. The resolution was computed out to the 14th term.
- The Sylow 2-subgroup of the Conway group Co_3 has order 2^{10}. The resolution was computed out to the 10th term.
- The Sylow 2-subgroup of the Mathieu group M_{24} has order 2^{10}. The resolution was computed out to the 9th term.

Structure of this book

The new two-speed version of Bergman's Diamond Lemma is demonstrated in Chap. 2. Prior to this the preferred basis $E_{\mathbf{X}}$ of a free kG-module is defined in Chap. 1 and negative word orderings are discussed. Then in Chap. 3 the variants of the Buchberger Algorithm are derived which lead to the method for constructing minimal resolutions.

The second part of the book is concerned with computing the cohomology ring from the miminal resolution. For this Gröbner bases for graded-commutative algebras are required, so they are introduced in Chap. 4.

Chap. 5 describes how to determine all generators and relations of the cohomology ring out to degree N given the first N terms of the minimal resolution. The methods are due to Carlson, but one has to consider how to carry them out using Gröbner bases. In order to keep the Gröbner basis of the relations ideal as small as possible, the monomial ordering and the method for choosing new generators are designed to work very closely together. Then Carlson's Completeness Criterion is recalled in Chap. 6 and its implementation is discussed briefly.

Finally Chap. 7 describes the results achieved to date with the package Diag. For space reasons the computed cohomology rings are not printed in full here. Instead they are available on the World Wide Web at the above address. Some salient samples are printed in the appendix.

Notation

Let p be a prime number and G a finite p-group. Cohomology groups $\mathrm{H}^n(G)$ and cohomology rings $\mathrm{H}^*(G)$ are always with coefficients in a field k of characteristic p.

Part I

Constructing minimal resolutions

1 Bases for finite-dimensional algebras and modules

Let G be a p-group and k a field of characteristic p. In the next chapter we shall construct a new kind of Gröbner bases for kG-modules. A prerequisite is that each free kG-module be assigned a preferred k-basis with certain properties. Calculations with the new Gröbner bases are performed by manipulating coordinate vectors with respect to the preferred k-basis.

This is unusual for Gröbner bases and simultaneously very interesting for computer calculations. For the space required to store a coordinate vector is known in advance and does not depend on its value. By contrast, Gröbner basis methods usually involve manipulating elements of a free k-algebra, and the space required to store such an element does of course depend on its value. There is a good reason for this: Gröbner bases were originally developed to perform calculations in infinite-dimensional algebras and their uses in the finite-dimensional case were only recognised later.

In this chapter each free kG-module F is assigned a preferred k-basis $E_{\mathbf{X}}(F)$. The most important case is the preferred k-basis $N_{\mathbf{X}}$ of the group algebra kG.

One k-basis for the group algebra is already known, namely the elements of the group. But as we want to work with projective resolutions it is better to choose a basis which contains a basis of the radical $\mathrm{rad}(kG)$. And since we want to work with Gröbner bases the preferred k-basis should be the set of the irreducible words for a presentation of the group and a suitable word ordering.

This word ordering need not be a well-ordering, despite the fact that this is usually necessary when working with Gröbner bases. For elements of the preferred basis can be multiplied using the group multiplication instead of by reduction over a Gröbner basis. So we may also use negative[1] word orderings, where the trivial word is the *largest* word. In fact the method used in Chap. 3 to determine minimal generators for a module only works for a negative word ordering.

Useable word orderings are introduced in Sect. 1.1 and the preferred k-basis $N_{\mathbf{X}}$ of the group algebra kG is defined. This depends on the chosen presentation of the group and on the useable ordering. The preferred k-basis $E_{\mathbf{X}}(F)$ of a free kG-module F is defined in Sect. 1.2. Then we discuss in Sect. 1.3 and

---

[1] In the sense of Mora [50]

Sect. 1.4 how to determine the preferred k-basis $N_\mathbf{X}$ and how to multiply elements of this basis.

The package Diag only uses the useable ordering $\leq_{\mathrm{RLL}}$ of Example 1.3. In Sect. 1.5 another useable ordering is presented which is constructed using the Jennings series of the group and which has certain interesting properties.

1.1 Finite-dimensional algebras

Definition 1.1. *Let* $\mathbf{X}$ *be a finite set. A useable ordering on the free monoid* $\langle \mathbf{X} \rangle$ *is a total ordering* $\leq$ *satisfying:*

1. *If* $u, w_1, w_2, v \in \langle \mathbf{X} \rangle$ *are words and* $w_1 \leq w_2$ *then* $uw_1v \leq uw_2v$.
2. *The set* $\{v \in \langle \mathbf{X} \rangle \mid v \geq w\}$ *is finite for each* $w \in \langle \mathbf{X} \rangle$.

Remark 1.2. Then 1 is the greatest element of $\langle \mathbf{X} \rangle$, and $\langle \mathbf{X} \rangle$ contains no infinite strictly decreasing sequences which are bounded below. So useable orderings are negative word orderings in the sense of Mora [50].

Example 1.3. (*The reverse length-lexicographical ordering* $\leq_{\mathrm{RLL}}$)
An ordering on $\mathbf{X}$ induces a lexicographical ordering $\leq_{\mathrm{lex}}$ on $\langle \mathbf{X} \rangle$. Denote by $\ell(w)$ the length of a word $w \in \langle \mathbf{X} \rangle$. The ordering $\leq_{\mathrm{RLL}}$ on $\langle \mathbf{X} \rangle$ is defined by

$$u \leq_{\mathrm{RLL}} v \quad \Longleftrightarrow \quad \ell(u) > \ell(v), \quad \text{or} \quad \ell(u) = \ell(v) \text{ and } u \geq_{\mathrm{lex}} v$$

for all $u, v \in \langle \mathbf{X} \rangle$. This is the ordering used in the package Diag. It is useable.

Example 1.4. Suppose that $\leq'$ is an ordering on $\langle \mathbf{X} \rangle$ respecting multiplication; and each $x \in \mathbf{X}$ has been assigned a positive integer as its dimension, inducing a dimension $\dim(w) \geq 0$ for each $w \in \langle \mathbf{X} \rangle$. Define an ordering $\leq$ on $\langle \mathbf{X} \rangle$ thus:

$$u \leq v \quad \Longleftrightarrow \quad \dim(u) > \dim(v), \quad \text{or} \quad \dim(u) = \dim(v) \text{ and } u \leq' v$$

for all $u, v \in \langle \mathbf{X} \rangle$. This is a useable ordering on $\langle \mathbf{X} \rangle$.

The ordering $\leq_{\mathrm{RLL}}$ is of this kind: $\leq'$ is the reverse lexicographical ordering and dimension coincides with length. In Sect. 1.5 we shall meet another ordering of this kind, but where some $x \in \langle \mathbf{X} \rangle$ have dimension greater than one.

Hypothesis 1.5. Let k be a field of characteristic p. Let Λ be a finite-dimensional k-algebra (associative with one) satisfying $\Lambda/\mathrm{rad}(\Lambda) \cong k$. Suppose we are given a presentation $0 \to I \longrightarrow k\langle \mathbf{X} \rangle \overset{\phi}{\longrightarrow} \Lambda \to 0$ of Λ with the following properties: The set $\mathbf{X}$ is finite and the two-sided ideal $J \subseteq k\langle \mathbf{X} \rangle$ generated by $\mathbf{X}$ is the inverse image of $\mathrm{rad}(\Lambda)$. Suppose further a useable ordering has been chosen on $\langle \mathbf{X} \rangle$.

Example 1.6. The case that we will consider in the later chapters is as follows: Let G be a finite p-group and $g_1, \ldots, g_m$ minimal generators for G. Then the group algebra $\Lambda := kG$ satisfies Hypothesis 1.5, where $\mathbf{X} := \{a_1, \ldots, a_m\}$ and

- The epimorphism $\phi\colon k\langle \mathbf{X}\rangle \to kG$ maps a_i to $g_i - 1$;
- The ordering on $\langle \mathbf{X}\rangle$ is the reverse length-lexicographical ordering $\leq_{\mathrm{RLL}}$. (Or more generally, any ordering from Example 1.4.)

Remark 1.7. Assuming Hypothesis 1.5, Nakayama's Lemma says there is an $n_I \geq 1$ satisfying $J^{n_I} \subseteq I \subseteq J$.

If A is an ordered set then every nonzero $f \in kA$ has a leading monomial $LM(f) := \max \operatorname{supp}(f)$. Write $LC(f)$ for the leading coefficient and $LT(f)$ for the leading term $LT(f) = LC(f)LM(f)$. Finally, for a subset T of kA set

$$LM(T) := \{LM(f) \mid f \in T \setminus \{0\}\}.$$

Lemma-Definition 1.8. *Assume Hypothesis 1.5. Define a subset $N_{\mathbf{X}} \subseteq \langle \mathbf{X}\rangle$ by $N_{\mathbf{X}} := \langle \mathbf{X}\rangle \setminus LM(I)$. Then $N_{\mathbf{X}}$ is finite and the map $kN_{\mathbf{X}} \hookrightarrow k\langle \mathbf{X}\rangle \twoheadrightarrow \Lambda$ is an isomorphism of k-vector spaces.*

Write $\mathcal{N}$ for the k-linear map $k\langle \mathbf{X}\rangle \to \Lambda \xrightarrow{\cong} kN_{\mathbf{X}}$ and $*$ for the map $kN_{\mathbf{X}} \times kN_{\mathbf{X}} \to kN_{\mathbf{X}}$ induced by the multiplication of Λ. Then $A*B = \mathcal{N}(AB)$ for $A, B \in kN_{\mathbf{X}}$, where AB denotes the product in $k\langle \mathbf{X}\rangle$.

Proof. Write ϕ for the quotient map $k\langle \mathbf{X}\rangle \twoheadrightarrow \Lambda$ and let n_I be as in Remark 1.7. Then the subset U of $\langle \mathbf{X}\rangle$ defined by $U := \{w \in \langle \mathbf{X}\rangle \mid \ell(W) < n_I\}$ is finite, and $\phi(w) = 0$ for every $w \in \langle \mathbf{X}\rangle - U$. Since $N_{\mathbf{X}}$ is clearly a subset of U, it is finite.

For $v \in \langle \mathbf{X}\rangle$ set $\langle \mathbf{X}\rangle_v := \{w \in \langle \mathbf{X}\rangle \mid w < v\}$ and $U_v := U \cap \langle \mathbf{X}\rangle_v$. Hence $\phi(k\langle \mathbf{X}\rangle_v) = \phi(kU_v)$ for each $v \in \langle \mathbf{X}\rangle$. So as $N_{\mathbf{X}} = \{w \in \langle \mathbf{X}\rangle \mid \phi(w) \notin \phi(k\langle \mathbf{X}\rangle_w)\}$ one has $N_{\mathbf{X}} = \{u \in U \mid \phi(u) \notin \phi(kU_u)\}$. Since U satisfies $\phi(kU) = \phi(k\langle \mathbf{X}\rangle)$ and is finite, the family $(\phi(w))_{w \in N_{\mathbf{X}}}$ is indeed a k-basis for Λ. $\qquad\square$

Example 1.9. Let G be the dihedral group $D_8 = \langle A, B \mid A^2, B^2, (AB)^4\rangle$ and k the field $\mathbb{F}_2$. Then A, B are minimal generators for G, and the corresponding presentation for kG is $kG \cong k\langle a, b\rangle / (a^2, b^2, abab + baba)$, where a maps to $A+1$ and b to $B+1$. In the ordering $\leq_{\mathrm{RLL}}$ one has $1 > a > b > a^2 > ab > ba > b^2 > a^3 > \cdots$. The preferred basis $N_{\mathbf{X}}$ is $\{1, a, b, ab, ba, aba, bab, baba\}$. The smallest generating set of the two-sided ideal $LM(I) \subseteq \langle \mathbf{X}\rangle$ is $\{a^2, b^2, abab\}$. Here are two examples of $*$-products: $a * (a + b) = \mathcal{N}(a^2 + ab) = ab$ and

$$baba * b = \mathcal{N}(baba.b) = \mathcal{N}(b.abab) = \mathcal{N}(b^2 aba) = 0.$$

Lemma 1.10. *Assume Hypothesis 1.5. Denote by S the smallest generating set of the right ideal $LM(I)$ in $\langle \mathbf{X}\rangle$, and let n_I be as in Remark 1.7. Then S is finite and*

$$\sum_{\sigma \in S} (\sigma - \mathcal{N}(\sigma))k\langle \mathbf{X}\rangle + J^{n_I} = I.$$

Proof. S is finite because $N_{\mathbf{X}}$ is. Denote by U the set

$$U := \{w \in \langle \mathbf{X} \rangle \mid \text{There is } v \in \langle \mathbf{X} \rangle \text{ with } \ell(v) < n_I \text{ and } v \leq w\}.$$

This is a finite set by the definition of useable ordering. Now let f be an element of I. By repeatedly subtracting expressions of the form $\lambda(\sigma - \mathcal{N}(\sigma))w$ with $\lambda \in k^*$, $\sigma \in S$ and $w \in \langle \mathbf{X} \rangle$ one arrives at either $f = 0$ or $LM(f) \notin U$. But then $\mathrm{supp}(f) \cap U = \emptyset$ and so $f \in J^{n_I}$. $\qquad\qquad\square$

1.2 Free right modules

Let $F := \bigoplus_{i=1}^m e_i \Lambda$ be a free right Λ-module.

Definition 1.11. *Set* $E(F) := \{e_i w \mid 1 \leq i \leq m, w \in \langle \mathbf{X} \rangle\}$ *and define the preferred k-basis* $E_{\mathbf{X}}(F)$ *of F by* $E_{\mathbf{X}} := \{e_i w \in E \mid w \in N_{\mathbf{X}}\}$. *Define maps* $w \colon E \to \langle \mathbf{X} \rangle$ *and* $\nu \colon E \to \{e_1, \ldots, e_m\}$ *by* $w(e_i w) := w$ *and* $\nu(e_i w) := e_i$.

Definition 1.12. *An ordering on $E(F)$ is called useable if for all A, A_1, A_2 in E and B, B_1, B_2 in $\langle \mathbf{X} \rangle$ one has:*

1. *If $\nu(A_1) = \nu(A_2)$ then: $A_1 \leq A_2 \iff w(A_1) \leq w(A_2)$.*
2. *If $A_1 \leq A_2$ then $A_1 B \leq A_2 B$.*
3. *If $B_1 \leq B_2$ then $AB_1 \leq AB_2$ (follows from 1).*

Example 1.13. Order $\langle \mathbf{X} \rangle$ with the ordering $\leq_{\mathrm{RLL}}$ and order $\{e_1, \ldots e_m\}$ by $e_i \leq e_j \iff i \leq j$. Then we may order E as follows: $A_1 < A_2$ if and only if

- $\ell(w(A_1)) > \ell(w(A_2))$; or
- Same length, and $\nu(A_1) < \nu(A_2)$; or
- Same length, $\nu(A_1) = \nu(A_2)$, and $w(A_1) >_{\mathrm{lex}} w(A_2)$.

Then $\leq$ is a useable ordering on E which is compatible with the ordering $\leq_{\mathrm{RLL}}$ on $\langle \mathbf{X} \rangle$. This is the ordering on E used in the package Diag.

Example 1.14. For $m = 2$ there is exactly one useable ordering which satisfies $e_1 B_1 > e_2 B_2$ for all $B_1, B_2 \in \langle \mathbf{X} \rangle$ and is compatible with $\leq_{\mathrm{RLL}}$ on $\langle \mathbf{X} \rangle$. This is an example of the elimination ordering we shall use later in Sect. 3.1.

Remark 1.15. Each (non-empty) subset of $\langle \mathbf{X} \rangle$ has a largest element, and so each (non-empty) subset of E has a largest element too. In particular, $\max(E)$ is one of the e_i.

By setting $\mathcal{N}(e_i w) := e_i \mathcal{N}(w)$ we extend $\mathcal{N}$ to a k-linear map $kE \to kE_{\mathbf{X}}$ satisfying $A * B = \mathcal{N}(AB)$ for all $A \in kE_{\mathbf{X}}$ and $B \in kN_{\mathbf{X}}$. Here, $*$ stands for the right Λ-action on F.

1.3 Implementation

Let $g_1, \ldots, g_m$ be minimal generators of the p-group G. As in Example 1.6 we may present kG as $k\langle \mathbf{X} \rangle / I$, where $\mathbf{X} = \{a_1, \ldots, a_m\}$ and a_i is mapped to $g_i - 1$.

The group algebra can be represented on the computer as $kN_{\mathbf{X}}$ together with the $*$-product. This representation is computed and stored as follows:

- The regular permutation representation of G:
  The group G may be specified as a permutation group or via a Power-Conjugate presentation. First one numbers the elements of G. This realises the regular permutation representation of G as a concrete embedding $G \rightarrowtail S_{|G|}$. The permutations in $S_{|G|}$ corresponding to $g_1, \ldots, g_m$ are stored, in the .reg file. From now on, this list of permutations is the definition of G.
- List the words in $N_{\mathbf{X}}$:
  We wrote ϕ for the epimorphism $k\langle \mathbf{X} \rangle \to kG$. The regular representation allows one to calculate the image $\phi(f) \in kG \cong k^{|G|}$ of any $f \in k\langle \mathbf{X} \rangle$. Also, one can compute the radical of any given ideal in kG.
  For a length r word $u \in \langle \mathbf{X} \rangle$ set $P_u := \{v \in \langle \mathbf{X} \rangle \mid \ell(v) = r$ and $v \leq_{\mathrm{RLL}} u\}$. Then $u \in N_{\mathbf{X}}$ if and only if $\phi(u)$ is linearly independent of $\phi(P_v)$ and $\mathrm{rad}^{r+1}(kG)$. The set $N_{\mathbf{X}}$ is stored as a list of words in the a_i, in decreasing order (.nontips file).
- The $*$-product on $kN_{\mathbf{X}}$:
  We now have two k-bases for kG, namely G and $\phi(N_{\mathbf{X}})$. Both basis change matrices are computed and stored (.bch file).
  For each i one first determines the matrix for right multiplication by $\phi(a_i)$ in kG with respect to the basis G. By change of basis one then obtains the matrix for right $*$-multiplication by a_i with respect to the basis $N_{\mathbf{X}}$. These matrices are stored (.gens file) and then used to compute the matrices for left $*$-multiplication by each a_i (.lgens file).

1.4 The matrix of a general element

Let f be an arbitrary element of the group algebra kG, let $R_f: kG \to kG$ denote the right action $h \mapsto h * f$, and $L_f: kG \to kG$ the left action $h \mapsto f * h$. We shall now see how to determine the matrix of L_f with respect to the basis $N_{\mathbf{X}}$ using the (previously computed) matrices of the R_a for $a \in \mathbf{X}$. Similarly one can calculate the matrix of R_f using the matrices of the L_a for $a \in \mathbf{X}$.

These matrices are needed when one has to compose two maps of free kG-modules. For this means multiplying two matrices with entries in kG, and that involves multiplying arbitrary elements of kG. But if f, h are elements of kG then $f * h = L_f(h) = R_h(f)$.

So let f be an element of kG and b an element of $N_{\mathbf{X}}$. If b is 1 then $L_f(b) = f * 1 = f$. If it is not, then there are $b' \in N_{\mathbf{X}}$ and $a \in \mathbf{X}$ satisfying

$b = b'.a$, and so $L_f(b) = f * (b'.a) = (f * b') * a = R_a(L_f(b'))$. Since R_a is known by assumption, we may calculate $L_f(b)$ given $L_f(b')$. So we may compute the matrix of L_f by induction on the length of b.

1.5 The Jennings ordering

In this section a useable ordering is constructed which has some interesting properties (see Proposition 1.22). Power-Conjugate presentations are important when working with p-groups on the computer. Jennings presentations and the Jennings ordering result on translating these methods from groups to group algebras. However this ordering is not currently used in the package Diag, as an efficient method for storing products during reduction in the Buchberger algorithm is lacking. A good storage method for the ordering $\leq_{\mathrm{RLL}}$ is described in Sect. 2.3.

For $r \geq 1$ the rth dimension subgroup $F_r(G)$ of a finite p-group G is defined by

$$F_r(G) := \{g \in G \mid g - 1 \in \mathrm{rad}^r(kG)\}.$$

Clearly $F_1 = G$ and $F_{r+1} \leq F_r$. Amongst others, the following properties of dimension subgroups are proved in [5, §3.14].

Proposition 1.16. *Let G be a finite p-group. The $F_r(G)$ form a decreasing central series. If r is large enough then $F_r = 1$. Moreover, $[F_r, F_s] \leq F_{r+s}$ and $\{g^p \mid g \in F_r\} \subseteq F_{pr}$ for all $r, s \geq 1$.* □

Note however that F_{r+1} can be equal to F_r without F_r being trivial. For example, if G is cyclic of order nine then F_2 and F_3 are both cyclic of order three.

Definition 1.17. *Let G be a group of order p^n. A system of Jennings PC-generators for G consists of elements $g_1, \ldots, g_n$ of G satisfying:*

- *If $g_i \in F_r$ then $g_{i+1} \in F_r$.*
- *The g_i with $g_i \in F_r - F_{r+1}$ are minimal generators for F_r/F_{r+1}.*

Jennings PC-generators do indeed lead to a polycyclic presentation for G:

Lemma 1.18. *Let $g_1, \ldots, g_n$ be Jennings PC-generators for G. Then g_i^p lies in $\langle g_{i+1}, \ldots, g_n \rangle$ for all $1 \leq i \leq n$ and $[g_i, g_j] \in \langle g_{j+1} \ldots, g_n \rangle$ for all $1 \leq i < j \leq n$.*

Proof. A consequence of Proposition 1.16. □

Example 1.19. Let $G = \langle a, b, c, \phi \rangle$ be the following semidirect product of order 32: The subgroup $\langle a, b, c \rangle$ is normal, and elementary abelian of order 8. The element ϕ has order 4, and its left conjugation action is $a \mapsto b \mapsto c \mapsto abc$.

Then $F_2 = \langle ab, ac, \phi^2 \rangle$ and $F_3 = \langle ac \rangle$. These are both elementary abelian. A system of Jennings PC-generators for G is a, ϕ, ab, ϕ^2, ac.

The last three of these generators are Jennings PC-generators for F_2. Another system of Jennings PC-generators for F_2 is ab, bc, ϕ^2, but note that $[\phi, bc] = ac$ and therefore this system cannot be extended to Jennings PC-generators for G.

Definition 1.20. *Let $g_1, \ldots, g_n$ be Jennings PC-generators for a p-group G.*

1. *Set $\mathbf{X} = \{a_1, \ldots, a_n\}$ and define an algebra epimorphism $\phi \colon k\langle \mathbf{X} \rangle \to kG$ by $a_i \mapsto g_i - 1$. Set $I = \mathrm{Ker}(\phi)$. Call $k\langle \mathbf{X} \rangle / I$ a Jennings presentation for kG.*
2. *Assign a_i dimension r if $g_i \in F_r - F_{r+1}$, so if $\phi(a_i) \in \mathrm{rad}^r(kG) - \mathrm{rad}^{r+1}(kG)$. So for each $w \in \langle \mathbf{X} \rangle$ the dimension $\dim(w)$ is at least as big as the length $\ell(w)$. Ordering $\mathbf{X}$ by $a_1 < a_2 < \cdots < a_n$ induces an ordering $\leq_{\mathrm{lex}}$ on $\langle \mathbf{X} \rangle$. The Jennings ordering $\leq_{\mathrm{J}}$ on $\langle \mathbf{X} \rangle$ is defined as follows:*

$$u \leq_{\mathrm{J}} v :\Longleftrightarrow \dim(u) > \dim(v), \quad or$$
$$\text{same dimension, and } \ell(u) < \ell(v), \quad or$$
$$\text{same dimension, same length, and } u \geq_{\mathrm{lex}} v.$$

Remark 1.21. The Jennings ordering is a useable ordering and belongs to the family of orderings constructed in Example 1.4. Together with its Jennings ordering, a Jennings presentation satisfies Hypothesis 1.5 for $\Lambda = kG$.

One advantage of the Jennings ordering is that the ideal $LM(I)$ in $\langle \mathbf{X} \rangle$ is particularly easy to determine:

Proposition 1.22. *Let $g_1, \ldots, g_n$ be a system of Jennings PC-generators for a p-group G, and $kG \cong k\langle \mathbf{X} \rangle / I$ the corresponding Jennings presentation. Then*

1. *The minimal generators of the two-sided ideal $LM(I)$ in $\langle \mathbf{X} \rangle$ are the words a_i^p for $1 \leq i \leq n$ and $a_i a_j$ for $1 \leq i < j \leq n$.*
2. *$N_{\mathbf{X}}$ consists of the words $a_n^{e_n} a_{n-1}^{e_{n-1}} \ldots a_1^{e_1}$ with $0 \leq e_i \leq p - 1$.*
3. *The set $\{\phi(w) \mid w \in N_{\mathbf{X}} \text{ and } \dim(w) = r\}$ is a k-basis for a complement of the subspace $\mathrm{rad}^{r+1}(kG)$ in $\mathrm{rad}^r(kG)$.*

Proof. We shall show that the words a_i^p and $a_i a_j$ (in the latter case for $i < j$) lie in $LM(I)$. The first two parts then follow immediately. Part 3. follows from 2. and Jennings' Theorem (Theorem 3.14.6 in [5]).

If $\dim(a_i) = r$ then $\phi(a_i^p)$ lies in kF_{pr}, which is generated by the $\phi(a_j)$ with $\dim(a_j) \geq pr$. So there is an $f \in k\langle \mathbf{X} \rangle$ such that $a_i^p - f \in I$ and each

$w \in \mathrm{supp}(f)$ satisfies either $\dim(w) > pr$ or $\ell(w) = 1$. Since a_i^p has length p, one has $a_i^p = LM(a_i^p - f)$.

For $i < j$ set $s := \dim(a_i)$, $t := \dim(a_j)$ and $c := [g_i, g_j]$, whence $c \in F_{s+t}$. Once more there is an $f \in k\langle \mathbf{X} \rangle$ such that $\phi(f) = c - 1$ and every $w \in \mathrm{supp}(f)$ satisfies $w < a_j a_i < a_i a_j$. Then $a_i a_j - a_j a_i - f - f a_j - f a_i - f a_j a_i$ lies in I and has leading monomial $a_i a_j$, because $g_i g_j = c g_j g_i$. $\qquad\square$

Example 1.23. Let G be the cyclic group of order four. A system of Jennings PC-generators for G is g, h with $g^2 = h$ and $h^2 = [g, h] = 1$. The corresponding Jennings presentation for $\mathbb{F}_2 G$ is $\mathbb{F}_2\langle a, b \rangle / (a^2 + b, ab + ba, b^2)$, with $\dim(a) = 1$ and $\dim(b) = 2$. The k-basis $N_{\mathbf{X}}$ is $\{1, a, b, ba\}$. In the Jennings ordering
$$1 > a > a^2 > b > ab > ba > a^4 > a^2 b > ba^2 > b^2 .$$

Remark 1.24. If G is not elementary abelian, then Jennings PC-generators for G cannot be minimal generators and so there are relations in $kG \cong k\langle \mathbf{X} \rangle / I$ which involve length one words. However the Jennings ordering ensures that each word in $LM(I)$ has length at least two.

2 The Buchberger Algorithm for modules

Let F be a free kG-module for a p-group G. In this chapter we derive a version of the Buchberger Algorithm which constructs Gröbner bases for submodules of F. Bergman's Diamond Lemma [12] is the template for the definitions and proofs of this chapter. In the next chapter we shall derive two variants of the Buchberger Algorithm which can be used to construct minimal projective resolutions.

There is a Diamond Lemma for modules in Bergman's paper [12, §9.5]. We shall here derive a new Diamond Lemma for modules which works throughout with a fixed k-basis for F, a fact which has considerable benefits for computer calculations. The new Diamond Lemma works by dividing Bergman's elementary reductions into two classes and only applying second class reductions to elements which are irreducible for the first class. The fixed k-basis for F is the basis $E_{\mathbf{X}}(F)$ of Chapter 1.

Here is a simple example of a two-speed reduction system:

Example 2.1. Set $\Lambda := \mathbb{F}_2\langle a \rangle / I$, where I is the two-sided ideal generated by $a + a^2$ and a^3. Clearly $I = (a)$.

We interpret the generators of I as a reduction system with two elementary reductions: r_1 replaces a by a^2, and r_2 replaces a^3 by 0. Then a is not reduction-finite, for the rule "always apply r_1 to the first factor" leads to

$$a \mapsto a^2 \mapsto a^3 \mapsto a^4 \mapsto \cdots.$$

By contrast if we declare r_2 to be first- and r_1 to be second-class then r_2 will always be used wherever possible. For example, we are forced to replace a^3 by 0 and may no longer apply r_1 to the first factor. So now every element of $\mathbb{F}_2\langle a \rangle$ is reduction-finite.

The core of this chapter is the Diamond Lemma for a two-speed reduction system (Theorem 2.12 in Section 2.1). This is used in Section 2.2 to derive the Buchberger Algorithm and in Section 2.3 we discuss implementation issues for this algorithm.

2.1 The Diamond Lemma for modules

We shall assume Condition 1.5 throughout, so Λ is a finite-dimensional k-algebra. As in §1.2 suppose further we are given a free right Λ-module $F = \bigoplus_{i=1}^{m} e_i \Lambda$ and a useable ordering $\leq$ on $E(F)$.

2.1.1 Reduction systems

Definition 2.2. *Let $M \subseteq F$ be a submodule. A reduction system for M is a family $f_T = (f_\tau)_{\tau \in T}$ satisfying*

1. *Each f_τ is an element of $kE_\mathbf{X} \setminus \{0\}$*
2. *$LC(f_\tau) = 1$ for all $\tau \in T$*
3. *Each f_τ lies in M.*

Each f_τ may therefore be written as $LM(f_\tau) - h_\tau$, where $h_\tau \in kE_\mathbf{X}$ and $A < LM(f_\tau)$ for each $A \in \mathrm{supp}(h_\tau)$.

Definition 2.3. *Call*

$$\mathrm{RS}(f_T) := \{(\tau, B) \in T \times N_\mathbf{X} \mid LM(f_\tau)B \in E_\mathbf{X}\}$$

the set of of T-reduction sites. To each such site (τ, B) associate an elementary T-reduction $r_{\tau B} \colon F \to F$, defined as the following k-linear map: $r_{\tau B}(A) = h_\tau * B$ for $A = LM(f_\tau)B$ and $r_{\tau B}(A) = A$ for $A \in E_\mathbf{X}$ otherwise. For $B = 1$ abbreviate $r_{\tau 1}$ to r_τ. A composition of finitely many elementary T-reductions shall be called a T-reduction.

So $r_{\tau B}(y) - y$ is a scalar multiple of $f_\tau * B$ for each $y \in F$. It follows that $r(y) - y \in M$ for every T-reduction r.

Definition 2.4. *Let y be an element of F.*

1. *y is irreducible if $r(y) = y$ for every T-reduction r. Denote by F_irr the k-vector subspace of irreducibles.*
2. *y is reduction-finite if the sequence $(r_i r_{i-1} \cdots r_1(y))$ is eventually constant for each sequence (r_i) of elementary reductions. If y is reduction-finite there are T-reductions r such that $r(y)$ is irreducible.*
3. *y is reduction-unique if it is reduction-finite and there is precisely one $z \in F_\mathrm{irr}$ such that there are T-reductions r satisfying $r(y) = z$. This element z is then denoted $r_T(y)$.*

Lemma 2.5. *Every element of F is reduction-finite.*

Proof. As $E_\mathbf{X}$ is an ordered finite set we may order its power set. This ordering is induced by the lexicographical ordering on the free monoid $\langle E_\mathbf{X} \rangle$, for we may view the power set of $E_\mathbf{X}$ as a subset of $\langle E_\mathbf{X} \rangle$: one turns subsets into words by writing down the elements of the subset in descending order.

Now let $r = r_{\tau B}$ be an elementary T-reduction. Then every element A of the support of h_τ is smaller than $LM(f_\tau)$ and so $AB < LM(f_\tau)B$. Moreover each element of the support of $A * B$ is at most AB. So $\operatorname{supp} r(y) \leq \operatorname{supp} y$, with equality if and only if $r(y) = y$. As the power set is finite, the support can only get smaller finitely many times. $\square$

Lemma 2.6. (Compare Lemma 1.1(i) in [12].) *The reduction-unique elements of F form a k-vector subspace, and the map r_T from this subspace to F_{irr} is k-linear.*

Proof. Let $y, z \in F$ be reduction-unique, let $\lambda, \mu \in k$, and let r be a T-reduction such that $r(\lambda y + \mu z)$ lies in F_{irr}. Then there are reductions r', r'' such that $r'r(y) = r_T(y)$ and $r''r'r(z) = r_T(z)$. Therefore $r(\lambda y + \mu z) = r''r'r(\lambda y + \mu z)$, which is just $\lambda r_T(y) + \mu r_T(z)$. $\square$

2.1.2 Ambiguities and the Diamond Lemma

Definition 2.7.   1. As in Lemma 1.10 denote by S the smallest generating set of $LM(I)$ as a right ideal in $\langle \mathbf{X} \rangle$. For $\sigma \in S$ set $h_\sigma := \mathcal{N}(\sigma) \in kN_{\mathbf{X}}$.
   2. Write M_T for $\sum_{\tau \in T} f_\tau * kN_{\mathbf{X}}$, the Λ-submodule of M generated by f_T. For $A \in E$ let T_A be the k-vector subspace of M spanned by

$$\{ f_\tau * B \mid \tau \in T,\ B \in N_{\mathbf{X}} \text{ and } LM(f_\tau)B < A \}.$$

Hence each T_A is a Λ-submodule of M_T.

Definition 2.8. Recall from Definition 1.11 that each $A \in E_{\mathbf{X}}$ is of the form $A = e_i w$, where $w \in N_{\mathbf{X}}$ and e_i is a free generator of F. We wrote $w(A) := w$.

1. An inclusion ambiguity is a 4-tuple (τ, t, A, B), where
   - τ, t are distinct elements of T;
   - B lies in $N_{\mathbf{X}}$ and A, AB in $E_{\mathbf{X}}$;
   - $LM(f_\tau) = A$ and $LM(f_t) = AB$.
   The ambiguity is called resolvable if there is a T-reduction r satisfying $r(f_t - f_\tau * B) = 0$. It is resolvable relative to $\leq$ if the S-polynomial $f_t - f_\tau * B$ lies in T_{AB}.
2. A toppling consists of a 4-tuple (τ, σ, A, B), where
   - τ lies in T and σ in S;
   - B lies in $N_{\mathbf{X}}$ and A in $E_{\mathbf{X}}$;
   - $LM(f_\tau) = A$, $\sigma = w(A)B$ and $w(A) \neq 1$.
   The toppling is resolvable if there is a T-reduction r satisfying $r(f_\tau * B) = 0$. It is resolvable relative to $\leq$ if the S-polynomial $f_\tau * B$ lies in T_{AB}. This toppling is called a toppling for τ.

Remark 2.9. Inclusion ambiguities are true ambiguities, for $r_{\tau B}$ and r_t are two distinct elementary T-reductions which alter AB. In contrast, topplings may be modelled on Bergman's overlap ambiguities, but they are not really ambiguities. This is because there is no elementary T-reduction $r_{\tau B}$. Rather, one has to replace AB by $A * B$ before one is allowed to apply T-reductions.

Remark 2.10. If (τ, σ, A, B) and (τ', σ', A', B') are two distinct topplings then $A'B'$ is not divisible by AB.

Example 2.11. Let G be the dihedral group D_8 and F the free right kG-module on one generator e. Recall (Example 1.9) that $k\langle a, b\rangle/(a^2, b^2, abab + baba)$ is a presentation of kG. Setting $f_{\tau_1} := eaba + ebab$ and $T := \{\tau_1\}$ yields a reduction system f_T without inclusion ambiguities. The set S has nine elements:

	σ_1	σ_2	σ_3	σ_4	σ_5	σ_6	σ_7	σ_8	σ_9
σ	a^2	b^2	ab^2	ba^2	aba^2	$abab$	bab^2	$baba^2$	$babab$
h_σ	0	0	0	0	0	$baba$	0	0	0

The topplings are $(\tau_1, \sigma_5, eaba, a)$ and $(\tau_1, \sigma_6, eaba, b)$ with S-polynomials $f_{\tau_1} * a = ebaba$ and $f_{\tau_1} * b = ebaba$ respectively. But $(\tau_1, \sigma_6, eaba, bab)$ is not a toppling, even though $eaba.bab = e.abab.ab = eab.abab$.

Theorem 2.12 (Diamond Lemma). (Compare Theorem 1.2 of [12].)
Let $f_T = (f_\tau)_{\tau \in T}$ be a reduction system for the submodule M of the right Λ-module F. Then the following are equivalent:

(a) All ambiguities (including topplings) are resolvable.
(a') All ambiguities (including topplings) are resolvable relative to $\leq$.
(b) All elements of F are reduction-unique and moreover

$$r_T(r(a) * b) = r_T(a * b)$$

for all $a \in kE_X$, $b \in kN_X$ and all T-reductions r.
(c) The irreducible $A \in E_X$ constitute a k-basis for the quotient module F/M_T.

If these statements hold then we may identify F/M_T with the k-vector space F_{irr}. For $a \in F_{\mathrm{irr}}$ and $b \in kN_X$ the Λ-action is given by $a \cdot b = r_T(a * b)$.

Remark 2.13. The second part of (b) corresponds to Lemma 1.1(ii) of [12]. The reason for this discrepancy is that Bergman's overlap ambiguities are true ambiguities whereas our topplings are not.

Proof (of Theorem 2.12). Clearly (a') follows from (a).

(c) $\Rightarrow$ (b): Let $y \in F$. Let r, r' be two T-reductions such that $r(y)$ and $r'(y)$ are irreducible. Then $r(y) - r'(y)$ lies in $M_T \cap F_{\mathrm{irr}} = \{0\}$ and so y is reduction-unique. Now suppose given r, a, b as in (b). Then $(r(a) - a) * b \in M_T$ and so the irreducible element $r_T(r(a) * b - a * b)$ lies in M_T and is therefore zero. By Lemma 2.6 r_T is linear.

(b) $\Rightarrow$ (a): Let (τ, σ, A, B) be a toppling. Then $h_\tau * B = r_\tau(A) * B$ and so $r_T(A * B) = r_T(h_\tau * B)$. Now let (τ, t, A, B) be an inclusion ambiguity. Then $h_\tau * B = r_{\tau B}(AB)$ and $h_t = r_t(AB)$, so $r_T(h_\tau * B) = r_T(h_t)$.

(b) $\Rightarrow$ (c): The map r_T is a projection from F to F_{irr} whose kernel is contained in M_T. Conversely M_T is generated as a Λ-module by the f_τ, and by (b) we have $r_T(f_\tau * b) = 0$ for all $b \in kN_{\mathbf{X}}$. So $\mathrm{Ker}(r_T) = M_T$.

(a') $\Rightarrow$ (b): Let $D \in E$. We shall prove the following statements:

1. Each $A \in E_{\mathbf{X}}$ with $A \leq D$ is reduction-unique.
2. $r_T(r(A) * B) = r_T(A * B)$ for all T-reductions r and all pairs (A, B) with $A \in E_{\mathbf{X}}$, $B \in N_{\mathbf{X}}$ and $AB \leq D$.

It suffices to prove the statements for every D in the finite set

$$\mathcal{D} := \{D \in E \mid D = AB \text{ for } A \in E_{\mathbf{X}}, B \in N_{\mathbf{X}}\}.$$

We shall do this by induction in $D \in \mathcal{D}$. The case $D = \max \mathcal{D} = \max E$ is the desired statement (b). For the inductive step (and to start the induction, if $D = \min \mathcal{D}$) we assume Statement 1 for $A < D$ and Statement 2 for $AB < D$.

Assume first that D lies in $E_{\mathbf{X}}$. If D is divisible by $LM(f_\tau)$ for at most one $\tau \in T$ then D is reduction-unique (inductive hypothesis for Statement 1). If (τ, t, A, B) is an inclusion ambiguity with $D = ABC$ for some $C \in N_{\mathbf{X}}$ then $r_{\tau BC}(D) = h_\tau * BC$ and $r_{tC}(D) = h_t * C$. Since $h_\tau * B - h_t = f_t - f_\tau * B$ lies in T_{AB} by (a'), the inductive hypothesis for Statement 2 implies $r_T(h_t * C) = r_T(h_\tau * B * C)$. So D is reduction-unique.

Now suppose given $A \in E_{\mathbf{X}}$ and $B \in N_{\mathbf{X}}$ with $D = AB$. It suffices to prove Statement 2 for elementary T-reductions. If $AB \in E_{\mathbf{X}}$ or $r(A) = A$ then this is trivial. So we may assume there are $\tau \in T$, $\sigma \in S$, $A' \in E_{\mathbf{X}}$ and $C_1, C_2, C_3 \in N_{\mathbf{X}}$ satisfying:

$$A = A'C_1, \quad B = C_2C_3, \quad LM(f_\tau) = A', \quad \sigma = w(A)C_2.$$

We must show that $r_T(r_{\tau C_1}(A) * B) = r_T(A * B)$, or equivalently that

$$r_T(f_\tau * C_1 * B) = 0. \tag{2.1}$$

We have two cases to consider:

- If $C_1C_2 \in N_{\mathbf{X}}$ then $(\tau, \sigma, A', C_1C_2)$ is a toppling. So (2.1) follows from (a') and the inductive hypothesis for Statement 2.
- If $C_1C_2 \in LM(I)$ then $C < C_1B$ for every $C \in \mathrm{supp}(C_1 * B)$. So (2.1) follows from the inductive hypothesis for Statement 2.

$\square$

Definition 2.14. *Let f_T be a reduction system for the submodule $M \subseteq F$ satisfying $M_T = M$ and the equivalent conditions of Theorem 2.12. Then f_T is a Gröbner basis for the module M. A Gröbner basis is minimal if it has no inclusion ambiguities.*

2.2 The Buchberger Algorithm

The Buchberger Algorithm Buchberger takes a finite generating set A for $M \subseteq F$ and transforms it into a minimal Gröbner basis. To do this Buchberger processes three finite lists:

- f_K is the current approximation to a Gröbner basis, a reduction system for M without inclusion ambiguities.
- The reduction system f_U contains elements of M waiting to be incorporated into f_K. The reduction system f_T generates M, where T is defined as $K \amalg U$.
- $\mathcal{X}$ contains the topplings of K which have yet to be made resolvable.

We shall need the function monic: $kE_{\mathbf{X}} \to kE_{\mathbf{X}}$ defined by monic$(0) = 0$ and monic$(y) = y/LC(y)$ for $y \neq 0$. Hence $LC(\text{monic}(y)) = 1$ for all $y \neq 0$.

An important component of Buchberger is Incorp. This algorithm incorporates the elements of f_U into f_K by removing inclusion ambiguities from f_T. At the same time $\mathcal{X}$ is updated accordingly.

Algorithm 2.15 (Incorp).
Processes lists $f_K, f_U, \mathcal{X}$ as above. At the end U is empty.

WHILE U is non-empty DO
    Choose a $u \in U$.
    IF f_T has an inclusion ambiguity (κ, u, A, B) with $\kappa \in K$ THEN
        $f_u := \text{monic}(f_u - f_\kappa * B)$
        IF $f_u = 0$ THEN $U := U \setminus \{u\}$ END IF
    ELSE
        FOR each inclusion ambiguity (u, κ, A, B) with $\kappa \in K$ DO
            Transfer κ from K to U.
            Remove every toppling for κ from $\mathcal{X}$.
        END FOR
        Transfer u from U to K.
        Append all topplings for u to $\mathcal{X}$.
    END IF
END WHILE

Proposition 2.16. *The algorithm Incorp stops in finite time. Let $T(0)$ be the initial state of T and $K(1)$ the final state of K. Then $M_{K(1)} = M_{T(0)}$, and $K(1)_A \supseteq T(0)_A$ for every $A \in E$. Moreover $f_{K(1)}$ has no inclusion ambiguities.*

We shall need one further definition for the proof of the proposition.

Definition 2.17. *For a reduction system f_T define*

$$VM(f_T) := \{A \in E_{\mathbf{X}} \mid A = LM(f_\tau)B \text{ with } \tau \in T, \, B \in N_{\mathbf{X}}\}.$$

Remark 2.18. *VM* stands for "Visible Monomials". There is a surjection from the set $RS(f_T)$ of T-reduction sites onto $VM(f_T)$ given by sending (τ, B) to $LM(f_\tau)B$. This surjection is a bijection if and only if the reduction system f_T has no inclusion ambiguities.

Proof (of Proposition 2.16). Each time a new element is added to K the set $VM(f_K)$ grows in size. As $E_{\mathbf{X}}$ is finite this can only happen finitely often.

To f_U we assign a word of length $|U|$ in the finite alphabet $E_{\mathbf{X}}$ by writing down the $LM(f_u)$ in ascending order. Call this word the signature of f_U. Consider what each iteration of the main loop in Incorp does to the signature. The size of U can increase if u is transferred from U to K, but this only happens finitely many times. If u is reduced because of an inclusion ambiguity then either $|U|$ becomes smaller or the signature decreases in the lexicographical ordering: that is, the signature decreases in the length-lexicographical ordering. As this is a well ordering the algorithm stops in finite time.

Now let (κ, u, A', B') be an inclusion ambiguity with $u \in U$ and $\kappa \in K$. If $f_u = f_\kappa$ then we may discard u without altering M_T or T_A. If they are distinct then there are $\lambda \in k \setminus \{0\}$ and $f'_u \in kE_{\mathbf{X}}$ satisfying $LC(f'_u) = 1$ and $f_u = f_\kappa * B' + \lambda f'_u$. The algorithm replaces f_u with f'_u. Since $LM(f'_u) < LM(f_u)$ there is no change to M_T, and T_A can only get larger.   □

The Buchberger Algorithm also requires the procedure Expand:

Procedure 2.19 (Expand).
Input: A toppling $\gamma = (\kappa, \sigma, A, B)$ in $\mathcal{X}$.

Remove γ from $\mathcal{X}$.
Append monic$(f_\kappa * B)$ to U if non-zero.

Algorithm 2.20 (Buchberger).
 Input: A finite reduction system f_U for M with $M_U = M$
Output: A minimal Gröbner basis f_K for M

$K := \emptyset$, $\mathcal{X} := \emptyset$
Carry out Incorp.
WHILE $\mathcal{X}$ is non-empty DO
     FOR each toppling γ in $\mathcal{X}$ DO
         Perform Expand on γ.
     END FOR
     Carry out Incorp.
END WHILE

The algorithm Buchberger is a special case of the following algorithm:

Algorithm 2.21.
Input: A finite reduction system f_V for M with $M_V = M$
Output: A minimal Gröbner basis f_K for M

$K := \emptyset,\, U := \emptyset,\, \mathcal{X} := \emptyset$
WHILE $V, \mathcal{X}$ not both empty DO
    EITHER
        Transfer a non-empty subset of V to U.
    OR
        Perform Expand on at least one element of $\mathcal{X}$.
    END EITHER
    Carry out Incorp.
END WHILE

Clarification 2.22. "EITHER ... OR ... END EITHER" means: perform exactly one of the two options. As $V, \mathcal{X}$ are not both empty at least one of the options can always be performed. If both can be performed, choose which one.

Theorem 2.23. *Algorithm 2.21 stops in finite time. The final state of f_K is a minimal Gröbner basis for M. Hence these statements also apply to the algorithm* Buchberger.

Proof. Each time a new element is added to K in Incorp, the set $VM(f_K)$ becomes larger. As $E_{\mathbf{X}}$ is finite, this can only happen finitely often, no matter how often Incorp is performed. Once K becomes constant, the list $\mathcal{X}$ shrinks each time Expand is performed and is left unchanged by Incorp. Similarly the list V shrinks until it becomes empty. So the algorithm stops.

Set $T := K \amalg U$ as above. Each time elements are transferred from V to U one has $T_A^{\text{old}} \subseteq T_A^{\text{new}}$ for every $A \in E$. By the time V becomes empty one has $M_T = M$.

Let K^1 be the state of K at the end of Algorithm 2.21. The rules governing $\mathcal{X}$ ensure that every toppling (κ, σ, A', B') with $\kappa \in K^1$ occurred in $\mathcal{X}$ at some point and then later as $f_\kappa * B'$ in U. At this later time the toppling is resolvable relative to $\le$ over f_T. Proposition 2.16 then says that the toppling is resolvable relative to $\le$ over K^1 too. The result follows from Theorem 2.12. $\qquad \square$

Remark 2.24. Algorithm 2.21 might appear unnecessarily general at first, but for actual calculations it has the advantage that it can be started when the list f_V is only partially known. If (as often happens) we know the value of $\dim_k(M)$ in advance then Lemma 2.25 tells us what the final size of $VM(f_K)$ should be. Once this size has been reached we may stop the algorithm, even if not all elements of f_V have yet been used.

Lemma 2.25. *Let f_T be a reduction system for the module $M \subseteq F$. Then:*

1. *f_T is a Gröbner basis for M if and only if $|VM(f_T)| = \dim_k(M)$.*
2. *f_T is a minimal Gröbner basis for M if and only if $|\mathrm{RS}(f_T)| = \dim_k(M)$ and f_T has no inclusion ambiguities.*

Proof. The statement $|VM(f_T)| = \dim_k(M)$ is equivalent to Statement (c) of the Diamond Lemma (Theorem 2.12). If there are no inclusion ambiguities then $|\mathrm{RS}(f_T)| = |VM(f_T)|$ by Remark 2.18. $\qquad\square$

2.3 Implementation

The Buchberger Algorithm 2.21 is not actually used in the package Diag. Rather it serves as a template for the algorithms ElimBuchberger (3.7) and HeadyBuchberger (3.18) which are used. But it is still worthwhile to consider how best to implement the Buchberger Algorithm, for here too the algorithm serves as a template for the two variants.

There are several places in the Buchberger Algorithm 2.21 and in its component Incorp (2.15) where more than one action is possible. For example when an element is to be chosen from a given list, or when one meets "EITHER ... OR ...". This generality helps in the proofs and means that the programmer has to choose from a plethora of possible strategies. The strategy used in the package Diag will now be described.

2.3.1 The algorithm Incorp

Experience shows that the most expensive part of the Buchberger Algorithm is calculating the products needed for reduction in Algorithm 2.15 (Incorp). For if (κ, u, A, B) is an inclusion ambiguity one needs to calculate the product $f_\kappa * B$ in order to determine the value of $r_{\kappa B}(f_u)$. Let p^n be the order of the group G, let m be the rank of the free kG-module F, and let ℓ be the length of the word $B \in N_{\mathbf{X}} \subseteq \langle \mathbf{X} \rangle$. The computer knows the $*$-product as the matrices (with respect to the basis $N_{\mathbf{X}}$) of the right $*$-action $f \mapsto f * a$ on kG for each $a \in \mathbf{X}$. So in order to calculate $f_\kappa * B$ the computer has to perform ℓm multiplications of a $p^n \times p^n$-matrix with a vector in $(\mathbb{F}_p)^{p^n}$.

Moreover the number of such elementary reductions that need to be calculated for one application of Incorp is typically very large, and an individual product $f_\kappa * B$ often gets used many times. It would take too long to recalculate $f_\kappa * B$ each time from scratch; and it would require too much space to compute and store all such products at the start of Incorp: the storage requirements would be comparable to constructing a k-basis of the module. But the reason Gröbner bases are being used is to handle cases where k-bases for the module are too big.

The following version of Incorp calculates each product $f_\kappa * B$ at most once, but also discards each product once it has been finished with and so manages to achieve acceptable space requirements.

Definition 2.26. *Let f_T be a reduction system for the submodule $M \subseteq F$. Order $\langle \mathbf{X} \rangle$ with the ordering $\leq_{\mathrm{RLL}}$ and $E(F)$ with the ordering of Example 1.13.*

1. *Define the dimension of $f \in kE(F) \setminus \{0\}$ with leading monomial $e_i w$ by*
   $$\dim(f) := \ell(w).$$
2. *For $d \geq 0$ define a set $\mathrm{RS}_d(f_T)$ and a family $\Pi_d(f_T)$ by*

$$\mathrm{RS}_d(f_T) := \{(\tau, B) \in T \times N_{\mathbf{X}} \mid \dim(f_\tau B) = d \text{ and } LM(f_\tau)B \in E_{\mathbf{X}}\},$$
$$\Pi_d(f_T) := (f_\tau * B)_{(\tau, B) \in \mathrm{RS}_d(f_T)}.$$

Algorithm 2.27. Order $E(F)$ as in Example 1.13.
Processes lists $f_K, f_U, \mathcal{X}$ as in Algorithm 2.15.

Write down $\Pi_0(f_K)$.
FOR $d \geq 0$ DO
    WHILE there are $u \in U$ with $\dim(f_u) = d$ DO
        Choose a $u \in U$ with $LM(f_u)$ as large as possible.
        IF f_T has an inclusion ambiguity (κ, u, A, B) with $\kappa \in K$ THEN
            Look up value of $f_\kappa * B$ in $\Pi_d(f_K)$.
            $f_u := \mathrm{monic}(f_u - f_\kappa * B)$
            IF $f_u = 0$ THEN $U := U \setminus \{u\}$ END IF
        ELSE
            FOR each inclusion ambiguity (u, κ, A, B) with $\kappa \in K$ DO
                Transfer κ from K to U.
                Remove each toppling for κ from $\mathcal{X}$.
            END FOR
            Transfer u from U to K.
            Append all topplings for u to $\mathcal{X}$.
            Append $(u, 1)$ to $\mathrm{RS}_d(f_K)$.
        END IF
    END WHILE
    IF U is empty THEN stop algorithm END IF
    Calculate $\Pi_{d+1}(f_K)$ from f_K and $\Pi_d(f_K)$.
    Discard $\Pi_d(f_K)$.
END FOR

Lemma 2.28. *Algorithm 2.27 is a special case of Algorithm 2.15.*

Proof. The ordering (1.13) on $E(F)$ satisfies $LM(f_1) > LM(f_2)$ if $\dim(f_1) < \dim(f_2)$. Let (κ, u, A, B) be an inclusion ambiguity with $f'_u \neq 0$ for $f'_u := \mathrm{monic}(f_u - f_\kappa * B)$. Then $LM(f'_u) < LM(f_u)$ and so $\dim(f'_u) \geq \dim(f_u)$.

    If $\dim(f_u) \geq d$ for every $u \in U$ at the beginning of the WHILE-loop then $\dim(f_u) = d$ for each u that is chosen. So if (κ, u, A, B) is an inclusion ambiguity then (κ, B) does lie in $\mathrm{RS}_d(f_K)$. Then at the end of the WHILE-loop $\dim(f_u) > d$ for every $u \in U$. So no elements of U are left behind. ⊔

Remark 2.29. Computing $\Pi_{d+1}(f_K)$ from f_K and $\Pi_d(f_K)$ is not difficult. Let (κ, B) be an element of $\mathrm{RS}_{d+1}(f_K)$, a set which is easy to determine. If $B = 1$ then $f_\kappa * B = f_\kappa$. Otherwise $B = B'a$ with $B' \in N_{\mathbf{X}}$, $a \in \mathbf{X}$ and

$(\kappa, B') \in \mathrm{RS}_d(f_K)$. Then $f_\kappa * B = (f_\kappa * B') * a$. The value of $f_\kappa * B'$ can be looked up in $\Pi_d(f_K)$ and so the value of $f_\kappa * B$ can be determined with $m := \mathrm{rank}(F)$ multiplications of a matrix with a vector.

Remark 2.30. If κ is transferred from K back to U then there is an inclusion ambiguity (u, κ, A, B) with $u \in U$ and $\dim(f_\kappa) \geq \dim(f_u)$ (equality will be possible in Algorithm 3.15). Hence κ is transferred back to U before any products $f_\kappa * w$ are calculated. In this sense no products are calculated unnecessarily.

Remark 2.31. If $\Pi_{d_0}(f_K)$ is known for some $d_0 \geq 0$ and $\dim(f_u) \geq d_0$ for all $u \in U$ then one can start Algorithm 2.27 with $d = d_0$ rather than $d = 0$. This often happens during the Buchberger Algorithm.

2.3.2 The Buchberger Algorithm

Definition 2.32. *Let f_T be a reduction system for the submodule $M \subseteq F$. Order $\langle \mathbf{X} \rangle$ with $\leq_{\mathrm{RLL}}$ and $E(F)$ with the ordering of Example 1.13.*

1. *A toppling $\gamma = (\tau, \sigma, A, B)$ has dimension $\dim(f_\tau B) = \ell(w(A)B)$.*
2. *For a list $\mathcal{X}$ of topplings set $\mathcal{X}_d := \{\gamma \in \mathcal{X} \mid \dim(\gamma) = d\}$.*

Algorithm 2.33. Order $E(F)$ as in Example 1.13.
Input: A finite reduction system f_V for M with $M_V = M$.
Output: A minimal Gröbner basis f_K for M.

$K := \emptyset$, $U := \emptyset$, $\mathcal{X} := \emptyset$, $d := 0$.
$\Pi_0(f_K)$ is the empty family.
WHILE $V, \mathcal{X}$ not both empty DO
    EITHER
        Discard $\Pi_d(f_K)$.
        Transfer a non-empty subset of V to U.
        $d := 0$.
        Write down $\Pi_0(f_K)$.
    OR
        Apply Expand to the elements of $\mathcal{X}_{d+1}$ using $\Pi_d(f_K)$.
        Calculate $\Pi_{d+1}(f_K)$ from f_K and $\Pi_d(f_K)$.
        Discard $\Pi_d(f_K)$.
        $d := d + 1$.
    END EITHER
    Carry out Incorp as in Algorithm 2.27.
END WHILE

Lemma 2.34. *Algorithm 2.33 is a special case of Algorithm 2.21.*

Proof. If $\gamma = (\kappa, \sigma, A, B)$ a toppling then $\dim(\gamma) \geq \dim(f_\kappa) + 1$. So no topplings get skipped over. $\qquad\square$

Remark 2.35. If $\gamma = (\kappa, \sigma, A, B)$ is a toppling with $\dim(\gamma) = d + 1$ then $B = B'a$ for $B' \in N_{\mathbf{X}}$ and $a \in \mathbf{X}$. So AB' in $E_{\mathbf{X}}$ since $\sigma = w(A)B$. Hence one may look up the value of $f_\kappa * B'$ in $\Pi_d(f_K)$ and compute $f_\kappa * B$ as $(f_\kappa * B') * a$.

Remark 2.36. Each time Incorp is carried out in Algorithm 2.33 one may start in dimension $d_0 := d$ instead of dimension 0 (cf. Remark 2.31).

Remark 2.37. No strategy has been given for the "EITHER ... OR ..." decision as here the Buchberger Algorithm is not a good model for the algorithms ElimBuchberger and HeadyBuchberger.

Remark 2.38. The dimension $\dim_k(M)$ is often known in advance, and then one can usually stop the Buchberger Algorithm early. For if $|\mathrm{RS}(f_K)| = \dim_k(M)$ at the end of Incorp then f_K is already a minimal Gröbner basis, by Lemma 2.25.

3 Constructing minimal resolutions

Minimal resolutions can be constructed by iterating the following key task:

> Given a homomorphism between two free kG-modules, compute *minimal generators* of the *kernel*.

In this chapter we will derive two variants of the Buchberger Algorithm that was developed in Chap. 2. Together these two variants perform the key task. The first variant is treated in Sect. 3.1: well-known elimination methods are used to determine the kernel. The second variant is developed in Sect. 3.2 and makes essential use of the fact that we are using a negative word ordering. A small alteration to the definition of an inclusion ambiguity ensures that the Buchberger Algorithm produces a minimal generating set as well as a Gröbner basis.

In Sect. 3.3 we shall consider pratical aspects of performing these two Buchberger variants. Finally we shall see in Sect. 3.4 how to use a by-product of the elimination to calculate preimages: we shall compute products in co-homology by lifting cocycles, and this calls for taking preimages.

3.1 The kernel of a homomorphism

Using the Buchberger Algorithm with an elimination ordering to determine the kernel of a map is a standard technique for commutative Gröbner bases. This method is also known in the noncommutative case [52].

Let $\phi: F_2 \to F_1$ be a Λ-linear map, where $F_1 = \bigoplus_{i=1}^{m} e_i \Lambda$ and $F_2 = \bigoplus_{j=1}^{n} \theta_j \Lambda$ are free right Λ-modules. Here, Λ is a finite dimensional k-algebra as in Hypothesis 1.5. Set $F := F_1 \oplus F_2$ and $\Gamma := \{\phi(f) - f \mid f \in F_2\}$. That is, Γ is the graph of $-\phi$. The following result is then clear:

Lemma 3.1. *As a Λ-module, Γ is generated by $\{\phi(\eta_j) - \eta_j \mid 1 \leq j \leq n\}$. Moreover $\Gamma \cap F_2 = \mathrm{Ker}(\phi)$.* □

Definition 3.2. *Suppose useable orderings on $E(F_1)$ and $E(F_2)$ have been chosen. There is exactly one ordering on $E(F) = E(F_1) \amalg E(F_2)$ which extends these two orderings and also satisfies $A_2 < A_1$ for all $A_1 \in E(F_1)$ and $A_2 \in E(F_2)$. This ordering is clearly useable. We shall call it an elimination ordering.*

Example 3.3. The elimination ordering used in the package Diag is obtained by choosing the ordering from Example 1.13 on $E(F_1)$ and $E(F_2)$.

Proposition 3.4. (Cf. [52, §4]) *Choose an elimination ordering for $E(F)$ and let f_T be a minimal Gröbner basis for Γ. Set $T_1 := \{\tau \in T \mid LM(f_\tau) \in F_1\}$ and $T_2 := T \setminus T_1$. Then*

1. *$T_2 = \{\tau \in T \mid f_\tau \in F_2\}$, and f_{T_2} is a minimal Gröbner basis for $\mathrm{Ker}(\phi)$.*
2. *$\{\pi_1(f_\tau) \mid \tau \in T_1\}$ is a minimal Gröbner basis for $\mathrm{Im}(\phi)$, where π_1 is projection $F \to F_1$.*
3. *Consider the reduction system f_{T_1} for Γ. Let f be an element of $\mathrm{Im}(\phi)$ and r a T_1-reduction such that $r(f)$ is irreducible over T_1. (Such reductions exist by Lemma 2.5.) Then $r(f)$ lies in F_2 and $\phi(r(f)) = f$.*

Proof. Let r_1 be a T_1-reduction and r_2 a T_2-reduction. Then $r_2(y) = y$ for each $y \in F_1$, and for each $y \in F_2$ we have $r_1(y) = y$ and $r_2(y) \in F_2$. Since f_T is a minimal Gröbner basis for Γ this implies (1) and (2).

Part (3): $r(f)$ lies in F_2 by (2). Now recall the definition of Γ.   □

Definition 3.5. *Because of properties (2) and (3) we shall call f_{T_1} a preimage Gröbner basis for $\mathrm{Im}(\phi)$. These properties will be useful when lifting coycles.*

The first variant ElimBuchberger of the Buchberger Algorithm computes a preimage Gröbner basis for $\mathrm{Im}(\phi)$ together with a minimal generating set for $\mathrm{Ker}(\phi)$. This algorithm and its component ElimIncorp process four finite lists:

- f_K is a reduction system for Γ without inclusion ambiguities. Moreover $K_1 = K$. The reduction system f_K is the current approximation to a preimage Gröbner basis for $\mathrm{Im}(\phi)$.
- The reduction system f_U contains elements of $\Gamma - F_2$ which are waiting to be incorporated into K. Set $T := K \amalg U$.
- f_V is a reduction system for $\mathrm{Ker}(\phi) = \Gamma \cap F_2$. At the end of ElimBuchberger f_V will be a generating set for the kernel. Γ is generated by the reduction set f_W, where W is defined as $K \amalg U \amalg V$.
- $\mathcal{X}$ contains the topplings of K which are not yet known to be resolvable over f_T.

Algorithm 3.6 (ElimIncorp).
Processes lists $f_K, f_U, \mathcal{X}, f_V$.
U is empty at the end.

WHILE U not empty DO
    Choose a $u \in U$.
    IF f_u lies in F_2 THEN
        Transfer u from U to V.
    ELSE IF f_T has an inclusion ambiguity (κ, u, A, B) with $\kappa \in K$ THEN
        $f_u := \mathrm{monic}(f_u - f_\kappa * B)$.

    IF $f_u = 0$ THEN $U := U \setminus \{u\}$ END IF
ELSE
        FOR each inclusion ambiguity (u, κ, A, B) with $\kappa \in K$ DO
            Transfer κ from K to U.
            Remove each toppling for κ from $\mathcal{X}$.
        END FOR
        Transfer u from U to K.
        Append all topplings for u to $\mathcal{X}$.
    END IF
END WHILE

*Algorithm 3.7 (*ElimBuchberger*).*
Input: $\phi(\eta_1), \ldots, \phi(\eta_n)$.
Output: Generating set f_V for $\text{Ker}(\phi)$ and preimage Gröbner basis f_K for $\text{Im}(\phi)$.

$f_U := \{\text{monic}(\phi(\eta_i) - \eta_i) \mid 1 \le i \le n\}$.
$K := \emptyset$, $\mathcal{X} := \emptyset$, $V := \emptyset$.
Carry out ElimIncorp.
WHILE $\mathcal{X}$ not empty DO
    Perform Expand on at least one element of $\mathcal{X}$.
    Carry out ElimIncorp.
END WHILE

Proposition 3.8. *The algorithm* ElimBuchberger *stops in finite time. When it stops f_V is a generating system for* $\text{Ker}(\phi)$ *and f_K is a preimage Gröbner basis for* $\text{Im}(\phi)$.

Proof. The elimination ordering ensures that ElimIncorp proceeds just like Incorp for $\text{Im}(\phi)$ with extra book-keeping. Similarly ElimBuchberger proceeds like Buchberger for $\text{Im}(\phi)$. So ElimIncorp and ElimBuchberger stop in finite time.

    Set $W := K \amalg U \amalg V$. Then f_W is a reduction system for Γ with $\Gamma_W = \Gamma$. Let $W^1 = K^1 \amalg V^1$ be the value of W at the end of ElimBuchberger and let (κ, σ, A, B) be a toppling of f_{W^1} with $\kappa \in K^1$. Immediately after it is expanded this toppling is resolvable relative to $\le$. As the vector space W_{AB} stays the same or becomes larger at each subsequent stage of ElimIncorp and ElimBuchberger, this toppling is also resolvable relative to $\le$ as a W^1-toppling. Applying the algorithm Buchberger to f_{V^1} produces a Gröbner basis f_{V^2} for Γ_{V^1}. Set $W^2 := K^1 \amalg V^2$. Then by the Diamond Lemma f_{W^2} is a minimal Gröbner basis for Γ, and K^1 is the set $\{\omega \in W \mid LM(f_\omega) \in F_1\}$. The result follows by Proposition 3.4. $\square$

3.2 Minimal generating sets

The Buchberger Algorithm can also be used to determine a minimal generating set of a module. For this it is necessary to distinguish between the elements of the reduction system which were there from the beginning and those which arise as expansions of topplings. For expansions lie in the radical of the module and therefore cannot belong to any minimal generating set. To record this distinction we shall use a *heady* reduction system.

Let M be a submodule of the free right Λ-module F, where Λ is a finite-dimensional k-algebra as in Hypothesis 1.5. Suppose a useable ordering has been chosen on $E(F)$.

3.2.1 Heady reduction systems

Definition 3.9. *A heady reduction system for M consists of a reduction system f_T for M together with a decomposition $T = T^H \amalg T^R$ such that f_ρ lies in $\sum_{\eta \in T^H} f_\eta * J$ for each $\rho \in T^R$.*

Recall that $J \subseteq k\langle \mathbf{X} \rangle$ is the inverse image of the radical $\mathrm{rad}(\Lambda)$.

Definition 3.10. *An elementary T-reduction $r_{\tau B}$ is called heady if $\tau \in T^H$ and $B = 1$, otherwise it is called radical. The composition of several radical elementary T-reductions is called a radical T-reduction.*

Definition 3.11. *For $A \in E$ define T_A^R to be the k-vector space spanned by*

$$\{ f_\tau * B \mid \tau \in T \text{ and } B \in N_{\mathbf{X}} \text{ with } LM(f_\tau)B < A, \text{ and } \tau \in T^R \text{ if } B = 1 \}.$$

A toppling (τ, σ, A, B) is called radically resolvable if there is a radical T-reduction r with $r(f_\tau * B) = 0$. It is called radically resolvable relative to $\leq$ if $f_\tau * B$ lies in T_{AB}^R.

Theorem 3.12. *Let M be a submodule of the free right Λ-module F and f_T a heady reduction system for M which has no inclusion ambiguities. Then the following statements are equivalent:*

(a) All topplings are radically resolvable.
(a′) All topplings are radically resolvable relative to $\leq$.
(b) All elements of F are reduction-unique. Moreover there is for each $a \in M_T * J$ a radical T-reduction r satisfying $r(a) = 0$.

If these equivalent statements hold then $\{ f_\eta \mid \eta \in T^H \}$ is a minimal generating set for M_T.

Proof. Clearly (a′) follows from (a).

(b) $\Rightarrow$ (a): If (τ, σ, A, B) is a toppling then $f_\tau * B$ lies in $M_T * J$. So there is a radical T-reduction r with $r(f_\tau * B) = 0$.

(a′) $\Rightarrow$ (b): The Diamond Lemma (Theorem 2.12) implies that all elements of F are reduction-unique. As before set

$$\mathcal{D} := \{ D \in E \mid D = AB \text{ for } A \in E_{\mathbf{X}}, B \in N_{\mathbf{X}} \}.$$

Each $a \in M_T * J$ can be written as

$$a = a_\lambda = \sum_{(\tau,B) \in T \times N_\mathbf{x}} \lambda_{\tau B} f_\tau * B$$

for suitable $\lambda_{\tau B} \in k$, where $\lambda_{\tau 1} = 0$ if $\tau \in T^H$. Define $D_\lambda \in \mathcal{D}$ by

$$D_\lambda := \max\{LM(f_\tau)B \mid \lambda_{\tau B} \neq 0\}.$$

The proof is by induction. Assume that for each λ with $D_\lambda < D$ there is a radical T-reduction r with $r(a_\lambda) = 0$. Now consider the case $D_\lambda = D$. Since there are no inclusion ambiguities there is exactly one pair (τ, B) with $LM(f_\tau)B = D$. If $D \in E_\mathbf{X}$ then $r_{\tau B}$ is a radical elementary T-reduction and $r_{\tau B}(a) = a_{\lambda'}$, where $\lambda'_{\tau B} = 0$ and $\lambda'_{tB'} = \lambda_{tB'}$ otherwise. Hence $D_{\lambda'} < D$ and the claim follows by the inductive hypothesis.
If D does not belong to $E_\mathbf{X}$ then there are $C_1, C_2 \in N_\mathbf{X}$ with $C_1 C_2 = B$ and a toppling $(\tau, \sigma, LM(f_\tau), C_1)$. By (a') we can write $f_\tau * C_1$ as a_μ with $D_\mu < LM(f_\tau)C_1$. Therefore we can write a as $a_{\lambda'}$ with $D_{\lambda'} < D$, and the claim follows.

By the definition of heady reduction system the family f_{T^H} generates M_T. Since there are no inclusion ambiguities the family $(LM(f_\eta))_{\eta \in T^H}$ is linearly independent over k and each k-linear combination is fixed by every radical T-reduction. So by (b) no nontrivial linear combination of f_{T^H} lies in $M_T * J$.

□

3.2.2 Algorithms for heady reduction systems

Let f_T be a heady reduction system.

Definition 3.13. *An inclusion ambiguity (τ, t, A, B) will be called inadmissible if $\tau \in T^H$, $t \in T^R$, and $B = 1$. All other inclusion ambiguities are admissible. Note that $LM(f_t) = LM(f_\tau)$ in the inadmissible case.*

Remark 3.14. Let (τ, t, A, B) be an inclusion ambiguity. If $t \in T^H$ then the ambiguity is admissible. If the ambiguity is inadmissible then $B = 1$ and $(t, \tau, A, 1)$ is an inclusion ambiguity which is admissible.

The Buchberger Algorithm Buchberger can be adapted to respect heady reduction systems. The new algorithm HeadyBuchberger again processes lists f_K, f_U und $\mathcal{X}$, but f_K, f_U and f_T are now heady reduction systems. Here, T is $K \amalg U$ and $T^R = K^R \amalg U^R$. Elements keep their type (R or H) on being transferred between lists. For example, when transferring from U to K, elements of U^R are sent to K^R and elements of U^H are sent to K^H.

Algorithm 3.15 (HeadyIncorp).
Processes lists $f_K, f_U, \mathcal{X}$.
At the end U is empty.

WHILE U not empty DO
    Choose a $u \in U$.
    IF f_T has an admissible inclusion ambiguity (κ, u, A, B) with $\kappa \in K$ THEN
        $f_u := \text{monic}(f_u - f_\kappa * B)$
        IF $f_u = 0$ THEN $U := U \setminus \{u\}$ END IF
    ELSE
        FOR each admissible inclusion ambiguity (u, κ, A, B) with $\kappa \in K$ DO
            Transfer κ from K to U.
            Remove each toppling for κ from $\mathcal{X}$.
        END FOR
        Transfer u from U to K.
        Append all topplings for u to $\mathcal{X}$.
    END IF
END WHILE

Proposition 3.16. HeadyIncorp *stops in finite time. Let* $f_{T(0)}$ *be the initial state and* $f_{K(1)}$ *the final state of the reduction system. Then* $M_{K(1)} = M_{T(0)}$ *and each* $A \in E$ *satisfies* $K(1)_A \supseteq T(0)_A$ *and* $K(1)_A^R \supseteq T(0)_A^R$. *Moreover* $f_{K(1)}$ *has no inclusion ambiguities.*

Proof. If (u, κ, A, B) is inadmissible then (κ, u, A, B) is admissible. So f_K stays free of inclusion ambiguities.

Each time a new element is added to K one of two things happens: either $VM(f_K)$ becomes larger or K^H becomes smaller. As $E_{\mathbf{X}}$ is finite this can only happen finitely many times.

We can form a word of length $|U|$ in the alphabet $E_{\mathbf{X}}$ by writing down the $LM(f_u)$ in increasing order. Only when a new element is added to K can $|U|$ increase. Each remaining step either makes $|U|$ smaller or makes the word associated to U larger in the lexicographical ordering. As there are only finitely many words of a given length, the algorithm does stop in finite time.

Now let (κ, u, A', B') be an admissible inclusion ambiguity with $u \in U$, $\kappa \in K$. If $f_u = f_\kappa$ then we may discard u without altering M_T, T_A, T_A^R. If they are not equal then there are $\lambda \in k \setminus \{0\}$ and $f_u' \in kE_{\mathbf{X}}$ such that $LC(f_u') = 1$ and $f_u = f_\kappa * B' + \lambda f_u'$. The algorithm replaces f_u with f_u'. As $LM(f_u')$ is smaller than $LM(f_u)$, there is no change to M_T, and T_A can only get larger. If $B' = 1$ and $\kappa \in K^H$ then u lies in U^H, for the ambiguity is admissible. So T_A^R too can only get larger. $\qquad\square$

*Procedure 3.17 (*HeadyExpand*).*
 Input: A toppling $\gamma = (\kappa, \sigma, A, B)$ in $\mathcal{X}$.

Remove γ from $\mathcal{X}$.
Append $\text{monic}(f_\kappa * B)$ to U^R if nonzero.

Now we can state the Buchberger Algorithm for heady reduction systems. This time we shall jump straight to the more general case.

*Algorithm 3.18 (*HeadyBuchberger*).*
Input: A finite reduction system f_V for M with $M_V = M$
Output: A minimal generating set f_{K^H} for M.

$K := \emptyset$, $U := \emptyset$, $\mathcal{X} := \emptyset$
WHILE $V, \mathcal{X}$ not both empty DO
    EITHER
        Transfer a nonempty subset of V to U^H
    OR
        Perform HeadyExpand on at least one element of $\mathcal{X}$.
    END EITHER
    Carry out HeadyIncorp.
END WHILE

Remark 3.19. Briefly, the modifications to the usual Buchberger Algorithm are as follows:

1. Some inclusion ambiguities are declared to be inadmissible.
2. Brand new generators are put in U^H.
3. Expansions are put in U^R.

Theorem 3.20. HeadyBuchberger *stops in finite time. At the end $\{f_\eta \mid \eta \in K^H\}$ is a minimal generating set for M.*

Proof. Each time a new element is added to K in HeadyIncorp, one of two things happens: either $VM(f_K)$ becomes larger or $|K^H|$ becomes smaller. In the latter case $VM(f_K)$ stays the same: it can never decrease in size. Since $E_{\mathbf{X}}$ is finite we conclude that the heady reduction system f_K is constant after finitely many applications of HeadyIncorp. After this, $\mathcal{X}$ shrinks each time HeadyExpand is performed and is left unchanged by HeadyIncorp. So after finite time the lists $\mathcal{X}$ and V are both empty and the algorithm stops.

Set $T := K \amalg U$ as above. Each time elements are transferred from V to U we have $T_A^{\text{old}} \subseteq T_A^{\text{new}}$ and $T^{\text{old}}{}_A^R \subseteq T^{\text{new}}{}_A^R$ for each $A \in E$. By the time V is empty we have $M_T = M$.

Let K^1 be the value of K at the end of HeadyBuchberger. The rules governing $\mathcal{X}$ ensure that each toppling (κ, σ, A', B') with $\kappa \in K^1$ occurred in $\mathcal{X}$ at some point and then later as $f_\kappa * B'$ in U^R. From this later time onwards the toppling is radically resolvable relative to $\leq$ over T. So Proposition 3.16 says that the toppling is radically resolvable relative to $\leq$ over K^1. The result follows from Theorem 3.12. $\qquad\qquad\square$

3.3 Implementation

Let $F_n = \bigoplus_{i=1}^{\beta(n)} kG$ be the nth term in the minimal projective resolution of the trivial right kG-module k. The nth differential $d_n: F_n \to F_{n-1}$ is characterised by the images in F_{n-1} of the $\beta(n)$ free generators of F_n. These images are stored in the .bin-file for d_n. The first differential d_1 is easy to write down. For $n \geq 1$ one calculates $\beta(n+1)$ and the images for d_{n+1} simultaneously by finding minimal generators for the kernel of d_n. For this the above algorithms are used, which only require as input the images under d_n of the free generators of F_n: they do not need the matrix of d_n as a k-linear map.

3.3.1 The variants of the algorithm Incorp

The algorithm EliminIncorp Let M be a submodule of the free module $F = F_1 \oplus F_2$. Impose on $E(F)$ the ordering of Example 3.3. Then $LM(f) \in E(F_1)$ for each $f \in kE(F) \setminus kE(F_2)$. For such f set $\dim(f) := \ell(w(LM(f)))$.

Let f_T be a reduction system for M. As in Proposition 3.4, define T_1, T_2 by $T_1 := \{\tau \in T \mid LM(f_\tau) \in E(F_1)\}$ and $T_2 := \{\tau \in T \mid f_\tau \in F_2\}$. By comparison with Definition 2.26 we then define a set $\mathrm{RS}_d(f_{T_1})$ and a family $\Pi_d(f_{T_1})$ by

$$\mathrm{RS}_d(f_{T_1}) := \{(\tau, B) \in T_1 \times N_{\mathbf{X}} \mid \dim(f_\tau B) = d \text{ and } LM(f_\tau)B \in E_{\mathbf{X}}(F)\}$$

and $\Pi_d(f_{T_1}) := (f_\tau * B)_{(\tau, B) \in \mathrm{RS}_d(f_{T_1})}$.

Algorithm 3.21 (Implementation of EliminIncorp).
Processes lists $f_K, f_U, \mathcal{X}, f_V$ as in Algorithm 3.6.
Order $E(F)$ as in Example 3.3.
Suppose given a $d_0 \geq 0$ such that $\dim(f_u) \geq d_0$ for each $u \in U_1$ and $\Pi_{d_0}(f_{K_1})$ is known. (Always true for $d_0 = 0$.)

Transfer all elements of U_2 to V.
FOR $d \geq d_0$ DO
    WHILE there are $u \in U = U_1$ with $\dim(f_u) = d$ DO
        Choose a $u \in U$ with $LM(f_u)$ as large as possible
        IF f_T has an inclusion ambiguity (κ, u, A, B) with $\kappa \in K$ THEN
            Look up value of $f_\kappa * B$ in $\Pi_d(f_{K_1})$.
            $f_u := \mathrm{monic}(f_u - f_\kappa * B)$
            IF $f_u = 0$ THEN
                $U := U \setminus \{u\}$
            ELSE IF f_u now lies in F_2 THEN
                Transfer u from U to V.
            END IF
        ELSE
            FOR each inclusion ambiguity (u, κ, A, B) with $\kappa \in K$ DO
                Transfer κ from K to U.
                Remove each toppling for κ from $\mathcal{X}$.
            END FOR
            Transfer u from U to K.
            Append $(u, 1)$ to $\mathrm{RS}_d(f_{K_1})$.
            Append all topplings for u to $\mathcal{X}$.

        END IF
    END WHILE
    IF U is empty THEN stop algorithm END IF
    Compute $\Pi_{d+1}(f_{K_1})$ from f_{K_1} and $\Pi_d(f_{K_1})$.
    Discard $\Pi_d(f_{K_1})$.
END FOR

The algorithm HeadyIncorp The implementation of this algorithm is closely modelled on the implementation of Incorp (cf. §2.3.1). We shall use the same notation.

Algorithm 3.22 (Implementation of HeadyIncorp*).*
Processes lists $f_K, f_U, \mathcal{X}$ as in Algorithm 3.15.
Order $E(F)$ as in Example 1.13.
Suppose given a $d_0 \geq 0$ such that $\dim(f_u) \geq d_0$ for each $u \in U$ and $\Pi_{d_0}(f_K)$ is known. (Always true for $d_0 = 0$.)

FOR $d \geq d_0$ DO
    WHILE there are $u \in U$ with $\dim(f_u) = d$ DO
        $A := \max\{LM(f_u) \mid u \in U\}$
        $V := \{u \in U \mid LM(f_u) = A\}$
        IF V^R is non-empty THEN
            Choose a $u \in V^R$.
        ELSE
            Choose a $u \in V^H$.
        END IF
        IF f_T has an admissible inclusion ambiguity (κ, u, A, B)
        with $\kappa \in K$ THEN
            Look up value of $f_\kappa * B$ in $\Pi_d(f_K)$.
            $f_u := \mathrm{monic}(f_u - f_\kappa * B)$
            IF $f_u = 0$ THEN $U := U \setminus \{u\}$ END IF
        ELSE
            FOR each admissible inclusion ambiguity (u, κ, A, B)
            with $\kappa \in K$ DO
                Transfer κ from K to U.
                Remove from $\mathcal{X}$ all topplings for κ.
                IF $LM(f_\kappa) = LM(f_u)$ THEN Remove $(\kappa, 1)$
                from $\mathrm{RS}_d(f_K)$. END IF
            END FOR
            Transfer u from U to K.
            Append to $\mathcal{X}$ all topplings for u.
            Append $(u, 1)$ to $\mathrm{RS}_d(f_K)$.
        END IF
    END WHILE
    IF U is empty THEN stop algorithm END IF
    Calculate $\Pi_{d+1}(f_K)$ from f_K and $\Pi_d(f_K)$.
    Discard $\Pi_d(f_K)$.
END FOR

3.3.2 The variants of the Buchberger Algorithm

The simplest strategy for determining the kernel of $\phi \colon F_2 \to F_1$ is the following algorithm. Here $\eta_1, \ldots, \eta_n$ are free generators for the free kG-module F_2.

Algorithm 3.23.
Input: $\phi(\eta_1), \ldots, \phi(\eta_n)$
Output: A minimal generating set f_A for $\mathrm{Ker}(\phi)$ and a preimage Gröbner basis f_B for $\mathrm{Im}(\phi)$.

Carry out ElimBuchberger on $\phi(\eta_1), \ldots, \phi(\eta_n)$.
$f_B :=$ the resulting preimage Gröbner basis f_K for $\mathrm{Im}(\phi)$.
Carry out HeadyBuchberger on the generating set f_V for $\mathrm{Ker}(\phi)$.
$f_A :=$ the resulting minimal generating set f_{K^μ} for $\mathrm{Ker}(\phi)$.

However the dimension of $\mathrm{Ker}(\phi)$ is often known in advance, for example when ϕ is a differential in the minimal resolution. We shall see that this knowledge can be used to make the algorithm HeadyBuchberger 3.18 stop earlier. Moreover[1] one can interweave the algorithms ElimBuchberger (3.7) and HeadyBuchberger and thus make the algorithm ElimBuchberger stop earlier too. This can lead to a considerable time saving.

We shall now assemble the composite algorithm, which requires five procedures and four conditions. First though we have to solve a problem of notation: we have given the indexing sets in ElimBuchberger and in Heady-Buchberger the same names as in the model algorithm Buchberger, namely K, U and $T := K \amalg U$. Rather than introduce further potentially confusing notation I consider it best to settle for the following convention:

> The reduction systems f_K, f_U and f_T in the two procedures derived from ElimBuchberger have nothing to do with the reduction systems f_K, f_U and f_T in the three procedures from HeadyBuchberger. The same goes for the two lists $\mathcal{X}$ of topplings.

It is the reduction system f_V which is responsible for communication between the two algorithms.

The five procedures will be stated first, then the algorithm itself, and then the four conditions.

*Procedure 3.24 (*InitElimBuchberger*).*
Input: $\phi(\eta_1), \ldots, \phi(\eta_n)$.

$f_U := \{\mathrm{monic}(\phi(\eta_i) - \eta_i) \mid 1 \leq i \leq n\}$.
$K :- \emptyset$, $\mathcal{X} := \emptyset$, $V := \emptyset$.
$\Pi_0(f_{K_1})$ is empty.
Carry out ElimIncorp as in in Algorithm 3.21, with $d_0 = 0$.
$\theta := 0$.

---

[1] As already hinted at in Remark 2.24.

Procedure 3.25 (LoopElimBuchberger).
Variables: as in InitElimBuchberger.
We have $\mathcal{X}_d = \emptyset$ for $d \leq \theta$.

Apply Expand to the elements of $\mathcal{X}_{\theta+1}$ using $\Pi_\theta(f_{K_1})$.
Calculate $\Pi_{\theta+1}(f_{K_1})$ from f_{K_1} and $\Pi_\theta(f_{K_1})$.
Discard $\Pi_\theta(f_{K_1})$.
$\theta := \theta + 1$.
Carry out ElimIncorp as in Algorithm 3.21, with $d_0 = \theta$.

Procedure 3.26 (InitHeadyBuchberger).
$K := \emptyset$, $U := \emptyset$, $\mathcal{X} := \emptyset$.
$\Pi_0(f_K)$ is empty.
$\rho := 0$.

Procedure 3.27 (ReadHeadyBuchberger).
Variables: as in InitHeadyBuchberger.
Input: New elements of $\mathrm{Ker}(\phi)$ in f_V.

Discard $\Pi_\rho(f_K)$.
Transfer all elements of V to U^H.
Write down $\Pi_0(f_K)$.
$\rho := 0$.
Carry out HeadyIncorp as in Algorithm 3.22, with $d_0 = 0$.

Procedure 3.28 (LoopHeadyBuchberger).
Variables: as in InitHeadyBuchberger.
We have $\mathcal{X}_d = \emptyset$ for $d \leq \rho$.

Apply HeadyExpand to the elements of $\mathcal{X}_{\rho+1}$ using $\Pi_\rho(f_K)$.
Calculate $\Pi_{\rho+1}(f_K)$ from f_K and $\Pi_\rho(f_K)$.
Discard $\Pi_\rho(f_K)$.
$\rho := \rho + 1$.
Carry out HeadyIncorp as in Algorithm 3.22, with $d_0 = \rho$.

Algorithm 3.29 (KernelInterwoven).
Input: $\nu := \dim_k(\mathrm{Ker}(\phi))$, and $\phi(\eta_1), \ldots, \phi(\eta_n)$.
Output: A minimal generating set f_A for $\mathrm{Ker}(\phi)$ and a preimage Gröbner basis f_B for $\mathrm{Im}(\phi)$.

Carry out InitElimBuchberger.
Carry out InitHeadyBuchberger.
REPEAT
    Carry out LoopElimBuchberger.
UNTIL Condition A is satisfied.
$f_B :=$ current value of f_K in ElimBuchberger.
Carry out ReadHeadyBuchberger.
WHILE Condition B not satisfied DO

     Carry out LoopHeadyBuchberger.
     IF Condition C is satisfied THEN
         REPEAT
             Carry out LoopElimBuchberger.
         UNTIL Condition D is satisfied.
         Carry out ReadHeadyBuchberger.
     END IF
END WHILE
$f_A :=$ current value of f_{K^H} in HeadyBuchberger.

Condition 3.30 (Condition A).
Meaning: Enough elements of $\mathrm{Ker}(\phi)$ have been found to make it worth starting HeadyBuchberger.
Condition: $|VM(f_{K_1})| = \dim_k(\mathrm{Im}(\phi))$ in ElimBuchberger, so f_{K_1} is a preimage Gröbner basis for $\mathrm{Im}(\phi)$. Moreover: either ElimBuchberger has finished (i.e., the list $\mathcal{X}$ is empty), or LoopElimBuchberger has been carried out at least n_A times since f_{K_1} first became a preimage Gröbner basis.
Explanation: Ideally we would stop ElimBuchberger as soon as f_V becomes large enough to generate $\mathrm{Ker}(\phi)$, but this condition is impractical to test on the computer. By contrast Condition A is easy to test, and experience shows that it is usually first satisfied shortly before or after f_V becomes a generating set for $\mathrm{Ker}(\phi)$. The parameter n_A allows for fine tuning and its optimal value seems to depend on the prime p. For $p = 2, 3$ a good value is $n_A = 2$. For $p = 5$ I currently use $n_A = 5$, for otherwise Condition A is often satisfied far too early. It is cheaper to have Condition A satisfied a little too late rather than a little too early.

Condition 3.31 (Condition B).
Meaning: The algorithm has found a minimal generating set for $\mathrm{Ker}(\phi)$ and can terminate.
Condition: $|VM(f_K)| = \dim_k(\mathrm{Ker}(\phi))$ in HeadyBuchberger. Also $U^H = \emptyset$, and: $\dim(f_\kappa) < \dim(f_u)$ and $\dim(f_\kappa) < \dim(\gamma)$ for all $\kappa \in K^H$, $u \in U$ and $\gamma \in \mathcal{X}$.
Explanation: By Lemma 2.25 f_K is a minimal Gröbner basis for $\mathrm{Ker}(\phi)$ if and only if $|VM(f_K)| = \dim_k(\mathrm{Ker}(\phi))$. The remaining conditions ensure that K^H remains constant for the rest of HeadyBuchberger, which means that the current state of f_{K^H} is also the final state. Note that the later parts of this condition are in particular satisfied if $U = \mathcal{X} = \emptyset$.

Condition 3.32 (Condition C).
Meaning: The elements of the kernel found up to now do not seem to generate. Suspend HeadyBuchberger and continue with ElimBuchberger.
Condition: $\mathcal{X} \neq \emptyset$ in ElimBuchberger and $|VM(f_K)| < \dim_k(\mathrm{Ker}(\phi))$ in HeadyBuchberger. Moreover either $\mathcal{X} = \emptyset$ in HeadyBuchberger, or the last n_C applications of LoopHeadyBuchberger have left $VM(f_K)$ unchanged.
Explanation: In practice this usually means that f_T does not generate $\mathrm{Ker}(\phi)$.

A good choice for $p = 2, 3$ seems to be $n_C = 2$. For $p = 5$ I use $n_C = 1$, as each cycle of LoopHeadyBuchberger takes a rather long time.

Condition 3.33 (Condition D).
Meaning: Suspend ElimBuchberger once more and continue with HeadyBuchberger.
Condition: $\mathcal{X}$ is now empty in ElimBuchberger; or new elements of $\mathrm{Ker}(\phi)$ have been found; or LoopElimBuchberger has now been performed n_D times without finding new elements of $\mathrm{Ker}(\phi)$.
Explanation: To check whether new elements of $\mathrm{Ker}(\phi)$ have been found, the computer performs ReadHeadyBuchberger at the end of LoopElimBuchberger.

The third part of the condition allows for the possibility that Condition C caused HeadyBuchberger to be suspended when in fact f_T did generate the kernel. Usually I set $n_D = 2$.

Lemma 3.34. *Algorithm 3.29* KernelInterwoven *stops in finite time. At the end f_A is a minimal generating set for* $\mathrm{Ker}(\phi)$ *and f_B is a preimage Gröbner basis for* $\mathrm{Im}(\phi)$.

Proof. All elements occurring in f_V belong to $\mathrm{Ker}(\phi)$, so f_A is a minimal generating set for the kernel provided the algorithm does stop. As ElimBuchberger stops in finite time, $\mathcal{X}$ is empty after finitely many applications of LoopElimBuchberger. So LoopElimBuchberger can only be performed finitely often and Condition C can only be satisfied finitely often. Once Condition C is never again satisfied, LoopHeadyBuchberger only has to be performed finitely often to make the $\mathcal{X}$ of HeadyBuchberger empty. But if $\mathcal{X}$ is empty in ElimBuchberger and in HeadyBuchberger, then Condition B will be satisfied. $\square$

3.4 Computing preimages

Let $d: F_2 \to F_1$ be a kG-linear map between free kG-modules, and let $(b_\rho)_{\rho \in R}$ be a finite family of elements of $\mathrm{Im}(d)$. Our task in this section is to compute a family $(u_\rho)_{\rho \in R}$ of elements of F_2 which satisfies $d(u_\rho) = b_\rho$ for all $\rho \in R$.

For example, let F_3 be another free kG-module and $\psi: F_3 \to F_1$ a kG-linear map whose image is contained in that of d. We would like to construct a lift $\overline{\psi}: F_3 \to F_1$ of ψ, that is a kG-linear map making the following diagram commute.

$$
\begin{array}{ccc}
F_3 & =\!=\!=\!= & F_3 \\
\overline{\psi} \downarrow & & \psi \downarrow \\
F_2 & \xrightarrow{\ d\ } & F_1
\end{array}
$$

Let $(e_\rho)_{\rho \in R}$ be free generators for F_3, set $b_\rho := \psi(e_\rho) \in \mathrm{Im}(d)$ and calculate preimages u_ρ of the b_ρ. Then a map $\overline{\psi}$ of the desired kind is constructed by setting $\overline{\psi}(e_\rho) := u_\rho$. This situation comes up when lifting cocycles to chain maps in order to compute the Yoneda product of two cohomology classes.

Tools for computing preimages were created in Sect. 3.1. Set $F := F_2 \oplus F_1$, choose an elimination ordering on $E(F)$ and use ElimBuchberger to calculate a preimage Gröbner basis for $\text{Im}(d)$.

Algorithm 3.35 (Lift).
Input: Family $(b_\rho)_{\rho \in R}$ in $\text{Im}(d)$, preimage Gröbner basis f_T for $\text{Im}(d)$.
Output: Preimages $(u_\rho)_{\rho \in R}$.

FOR each $\rho \in R$ DO $u_\rho := b_\rho$ END FOR
WHILE there are $\rho \in R$ with $LM(u_\rho)$ in F_1 DO
    Find $\tau \in T$ and $B \in N_\mathbf{X}$ satisfying $LM(f_\tau)B = LM(u_\rho)$.
    $u_\rho := u_\rho - LC(u_\rho).f_\tau * B$.
END WHILE

Proposition 3.36. *The algorithm* Lift *stops in finite time. At the end each* u_ρ *lies in* F_2 *and satisfies* $d(u_\rho) = b_\rho$.

Proof. By part 3 of Proposition 3.4, such τ, B can always be found. The result follows. $\qquad\qquad\qquad\qquad\qquad\qquad\qquad\qquad\qquad\qquad\qquad\qquad\quad$ $\square$

Now we shall state a version of Algorithm 3.35 Lift suitable for implementation. As in Sect. 3.1, for a reduction system f_R set

$$R_1 := \{\rho \in R \mid u_\rho \in F \setminus F_2\}.$$

Algorithm 3.37 (Implementation of Lift).
Input: Family $(b_\rho)_{\rho \in R}$ in $\text{Im}(d)$, preimage Gröbner basis f_T for $\text{Im}(d)$.
Output: Preimages $(u_\rho)_{\rho \in R}$.
Order $E(F)$ as in Example 3.3.
Observe that $T_1 = T$.

FOR each $\rho \in R$ DO $u_\rho := b_\rho$ END FOR
Write down $\Pi_0(f_T)$.
FOR $d \geq 0$ DO
    WHILE there are $\rho \in R_1$ with $\dim(u_\rho) = d$ DO
        Choose a $\rho \in R_1$ with $LM(f_\rho)$ as large as possible.
        Find $\tau \in T$ and $B \in N_\mathbf{X}$ which satisfy $LM(f_\tau)B = LM(u_\rho)$.
        $u_\rho := u_\rho - LC(u_\rho).f_\tau * B$
    END WHILE
    IF R_1 is empty THEN stop algorithm END IF
    Calculate $\Pi_{d+1}(f_T)$ from f_T and $\Pi_d(f_T)$.
    Discard $\Pi_d(f_T)$.
END FOR

Part II

Cohomology ring structure

4 Gröbner bases for graded commutative algebras

Let p be an odd prime and k a field of characteristic p. Cohomology rings over k are not commutative but rather graded commutative. So if x and y are homogeneous cohomology classes in degrees n and m respectively, then one has $y.x = (-1)^{nm} x.y$ rather than $y.x = x.y$. Consequently free graded commutative algebras are not in general polynomial algebras. So as we want to work with presentations of cohomology rings, we will need Gröbner bases for graded commutative rings. I know of no such theory in the literature, and so one will be developed here. Note that there is more than one possibility for such a theory.

If y is an element of a graded commutative algebra which is homogeneous of odd degree, then the relation $y^2 = 0$ is automatic. Relations of this kind hold even in *free* graded commutative algebras. This leads one to suspect that they will enjoy a privileged status in the desired Gröbner basis theory. The Gröbner bases that we shall now develop were designed with the aim of abolishing as far as possible the privileged status of these "structural" relations, in order to make the resultant methods easier to implement.

This principle that the Gröbner basis methods should be easy to implement actually leads to a different kind of Gröbner bases: not Gröbner bases for ideals in graded commutative algebras but rather Gröbner bases for right ideals in certain algebras which I shall call Θ-algebras. A Θ-algebra is what you get when you take the presentation as an associative algebra of a graded commutative algebra and then remove the relation $y^2 = 0$ for each odd dimensional free generator.

One sees from this definition of a Θ-algebra that a Θ-algebra has the same underlying vector space as the polynomial algebra on the same generators, but a different multiplication. In particular the bases of monomials is only closed up to sign under multiplication in the Θ-algebra. This means that we cannot use E. Green's Gröbner bases for algebras with a multiplicative basis [41, 42].

The Gröbner bases we shall develop for right ideals in Θ-algebras bear a strong resemblance to commutative Gröbner bases. A good text on these, the most widely known Gröbner bases, is Chapter 15 of Eisenbud's book [31]. The main difference is that at the start of the Buchberger Algorithm one has explicitly to add the element y^2 to the generating set of the right ideal for each odd dimensional generator y of the Θ-algebra.

Section 4.1 starts with the definition of a Θ-algebra and proceeds to establish some results on the structure of such algebras which will be needed to set up Gröbner bases for them. For example, right ideals satisfy the ascending chain condition (Proposition 4.16), every right ideal of interest to us is a two-sided ideal (Proposition 4.18), and any two monomials have a highest common factor and a lowest common multiple (Lemma 4.22).

Gröbner bases for right ideals in Θ-algebras are introduced in Sect. 4.2. As usual there is a Buchberger Algorithm for constructing Gröbner bases. The remaining sections contain applications: Sect. 4.3 describes how to calculate the kernel of a homomorphism of graded commutative algebras, and Sect. 4.4 introduces Gröbner bases for right modules over a Θ-algebra. These can be used to determine the intersection of two ideals and the annihilator of an element.

Remark 4.1. There is more than one way to set up Gröbner bases for graded commutative algebras. Here are some advantages of the current approach:

- As far as possible, the "structural" relations $y^2 = 0$ are treated in exactly the same way as the remaining relations. This makes it easier to list the critical pairs[1].
- Elements of a Θ-algebra may be represented as polynomials. The polynomial is unique, and every polynomial can occur.
- Gröbner bases for right ideals in a Θ-algebra are analogous to the usual commutative Gröbner bases. Gröbner bases for two-sided ideals would be more complicated as they would share at least some of the characteristics of noncommutative Gröbner bases.

Remark 4.2. If $p = 2$ then the cohomology algebra $\mathrm{H}^*(G)$ is a commutative k-algebra. So in this case one can use commutative Gröbner bases. A good source is Chapter 15 of Eisenbud's book [31].

Remark 4.3. The Gröbner bases we develop here have nothing to do with the Gröbner bases of Chap. 2. The Gröbner bases of that chapter are for modules over the group algebra of a p-group and are used to construct minimal resolutions. The Gröbner bases of the current chapter are for right ideals in Θ-algebras and are required in order to be able to compute in graded commutative algebras such as cohomology rings.

4.1 The structure of Θ-algebras

Definition 4.4. *Let k be a field of odd characteristic p. Let $z_1, \ldots, z_n$ be indeterminates, each equipped with a degree $|z_i| = t_i > 0$.*

---

[1] Pairs of relations whose leading monomials have a common factor.

1. The Θ-algebra $A = \Theta(z_1, \ldots, z_n)$ on z_i over k is defined as the associative algebra

$$\Theta(z_1, \ldots, z_n) := k\langle z_1, \ldots, z_n \rangle / (z_j z_i - (-1)^{t_i t_j} z_i z_j \text{ for } 1 \leq i < j \leq n).$$

The ordered n-tuple $(z_1, \ldots, z_n)$ is part of the structure of the Θ-algebra A.

2. $P(A) := k[z_1, \ldots, z_n]$ is called the associated polynomial algebra of the Θ-algebra A. We shall denote multiplication in $P(A)$ by $*_P$ to distinguish it from multiplication in A.

3. The family $\mathcal{B} = \mathcal{B}(A)$ of elements of A is defined as follows:

$$\mathcal{B} := \{z_1^{m_1} \cdots z_n^{m_n} \mid (m_1, \ldots, m_n) \in \mathbb{N}^n\}.$$

The following lemma hardly requires a proof.

Lemma 4.5. *Let $A = \Theta(z_1, \ldots, z_n)$ be a Θ-algebra.*

1. *The family $\mathcal{B} = \mathcal{B}(A)$ is k-basis of A.*
2. *$\mathcal{B}$ is a k-basis of the associated polynomial algebra $P(A)$ and closed under multiplication $*_P$ in the polynomial algebra.*
3. *For every pair $b_1, b_2 \in \mathcal{B}$ there is exactly one $\varepsilon_{12} \in \{-1, 1\}$ satisfying $b_1 b_2 = \varepsilon_{12} b_1 *_P b_2$.*

Proof. Order the monomials of $k\langle z_1, \ldots, z_n \rangle$ with the length-lexicographical ordering induced by $z_n > \cdots > z_1$. The reduction system

$$\{z_j z_i - (-1)^{t_i t_j} z_i z_j \mid 1 \leq i < j \leq n\} \qquad (t_i = |z_i|)$$

has no inclusion ambiguities and all overlap ambiguities are resolvable. So the Diamond Lemma [12] tells us that $\mathcal{B}$ is a basis. The rest is then clear. $\square$

Example 4.6. Let A be the algebra $\Theta(y, z)$ with y and z one-dimensional. Then $yz.y = -y^2 z$ and $yz *_P y = y^2 z$. Hence in this case we have $yz.y = -yz *_P y$.

Remark 4.7. More generally, if w is an arbitrary word in the generators of a Θ-algebra A then there are unique $b_w \in \mathcal{B}$ and $\varepsilon_w \in \{-1, 1\}$ satisfying $w = \varepsilon_w b_w$. For any two words v, w one has $vw = \pm wv$.

Remark 4.8. In general though, elements of a Θ-algebra do not commute, even up to a sign. For example, if y, z are both 1-dimensional then neither $(y + z)y = y(y + z)$ nor $(y + z)y = -y(y + z)$ holds in $\Theta(y, z)$.

Definition 4.9. *Let A be a Θ-algebra. A total ordering $\leq$ on $\mathcal{B}(A)$ shall be called useable if for all $b, b_1, b_2 \in \mathcal{B}$ it satisfies*

$$\text{If } b_1 \leq b_2 \text{ then } b_1 *_P b \leq b_2 *_P b.$$

Useable orderings exist for every Θ-algebra.

Example 4.10. The graded lexicographical ordering $\leq_{\text{grlex}}$ is defined by

$$b_1 \leq_{\text{grlex}} b_2 \Leftrightarrow \deg(b_1) < \deg(b_2), \text{ or } \deg(b_1) = \deg(b_2) \text{ and } b_1 \leq_{\text{lex}} b_2$$

This ordering is useable.

Remark 4.11. Let A be a Θ-algebra. By definition A is finitely generated and each generator is in positive degree. Therefore:

1. $\mathcal{B}(A)$ has a smallest element for the ordering $\leq_{\text{grlex}}$, namely 1.
2. The set $\{b \in \mathcal{B} \mid b \leq_{\text{grlex}} b_0\}$ is finite for each $b_0 \in \mathcal{B}$, as $\{b \in \mathcal{B} \mid |b| \leq n\}$ is finite for each $n \geq 0$,
3. $\leq_{\text{grlex}}$ is a well ordering.

Remark 4.12. Each element a of the Θ-algebra A has a support $\text{supp}(a) \subseteq \mathcal{B}$. If a useable ordering has been chosen then each $a \neq 0$ has a leading monomial $LM(a) \in \mathcal{B}$.

Lemma 4.13. *Let a_1, a_2 be elements of a Θ-algebra A. If $a_1 a_2 = 0$ then at least one of a_1, a_2 must be zero. If $a_1 a_2$ is a term (that is, its support has size one), then both a_1 and a_2 are terms.*

Proof. Choose a useable ordering and assume that a_1, a_2 are both nonzero. For $i = 1, 2$ let $\lambda_i b_i$ be the leading term of a_i. The product of these two terms is $\pm \lambda_1 \lambda_2 b_1 *_P b_2$, and no other product of a term in a_1 with a term in a_2 can produce a scalar mutliple of $b_1 *_P b_2$. The analogous statement holds for the product of the two smallest terms.   □

Definition 4.14. *A monomial ideal in the Θ-algebra A is an ideal which is generated by a subset of the basis $\mathcal{B}(A)$.*

Remark 4.15. Here we need not distinguish between left, right and two-sided ideals, because the left and the right ideal generated by a subset of $\mathcal{B}$ coincide.

Proposition 4.16. *Right ideals in a Θ-algebra satisfy the ascending chain condition.*

Proof. Let I be a right ideal in the Θ-algebra A. Order $\mathcal{B}(A)$ with the graded-lexicographical ordering, set

$$LM(I) = \{LM(f) \mid f \in I \setminus \{0\}\} \subseteq \mathcal{B}$$

and write $LT(I)$ for the subspace of A with k-basis $LM(I)$. Then $LT(I)$ is a monomial ideal in A. As $\mathcal{B}$ is also a k-basis of the associated polynomial algebra $P(A)$, we may also view $LT(I)$ as a monomial ideal in $P(A)$. Since $P(A)$ is noetherian there is a finite subset $S \subseteq LM(I) \subseteq \mathcal{B}$ which generates $LT(I)$ as an ideal in $P(A)$. Then S also generates $LM(I)$ as a right ideal in A.

Now choose for each $s \in S$ an $f_s \in I$ with $LT(f_s) = 1.s$. Denote by J the right ideal in A generated by the finite set $\{f_s \mid s \in S\}$. Clearly $J \subseteq I$. If J is not equal to I then there is an $f \in I \setminus J$ such that $LM(f) \leq_{\mathrm{grlex}} LM(g)$ for every $g \in f + J$, for $\leq_{\mathrm{grlex}}$ is a well-ordering. As however the term $LT(f)$ lies in $LT(I)$ there are $s \in S$ and a term t satisfying $LT(f) = st$. But this leads to a contradiction, because then $g := f - f_s t$ lies in $f + J$ and satisfies $LM(g) <_{\mathrm{grlex}} LM(f)$. $\qquad\square$

Remark 4.17. If $x_1, \ldots, x_n$ are even- and $y_1, \ldots, y_m$ odd-dimensional then the free graded commutative algebra on the x_i and the y_j is the quotient of the Θ-algebra on the same generators by the relations $y_i^2 = 0$ for $1 \leq i \leq m$.

Proposition 4.18. *Let A be the Θ-algebra $\Theta(x_1, \ldots, x_r, y_1, \ldots, y_s)$, where each x_i is even- and each y_j odd-dimensional. Let I be a homogeneous right ideal in A. If $y_j^2 \in I$ for every $1 \leq j \leq s$, then I is a two-sided ideal.*

The proof of the proposition requires a special case of the following lemma. Later on in Theorem 4.31 we shall use the lemma in its full generality.

Lemma 4.19. *Let A be the Θ-algebra $\Theta(x_1, \ldots, x_r, y_1, \ldots, y_s)$, where each x_i is even- and each y_j odd-dimensional. Let $f_1, \ldots, f_t$ be elements of A which are homogeneous of degree n, and $g_1, \ldots, g_t$ elements of A which are homogeneous of degree m. Set $\varepsilon := (-1)^{mn}$ and define*

$$\Phi := \sum_{i=1}^{t} g_i f_i - \varepsilon \sum_{i=1}^{t} f_i g_i \,.$$

Then there are elements $h_1, \ldots, h_s$ of A with the following properties:

1. *Each h_j is either zero or homogeneous of degree $n + m - 2$.*
2. *$\Phi = \sum_{j=1}^{s} y_j^2 h_j$.*
3. *$supp(y_j^2 h_j) \subseteq supp(\Phi)$ for each j.*

Proof. If b, b' are monomials of degree n, m which do not satisfy $b'b = \varepsilon bb'$ then $\gcd(b, b')$ involves at least one of the y_j. $\qquad\square$

Proof (of Proposition 4.18). For each homogeneous $f \in A$ and for each generator z of A, Lemma 4.19 implies that

$$zf - (-1)^{\mu\nu} fz \text{ lies in } I. \tag{4.1}$$

Here μ is the degree of f and ν that of z. Now let b be an element of $\mathcal{B}$. We shall prove by induction on the degree of b that bI is a subset of I. If b is a generator of A then the result follows from (4.1). If it is not then there are $b_1, b_2 \in \mathcal{B}$ satisfying $b = \pm b_1 b_2$ and both of positive degree. As their degrees are less than that of b we deduce that $b_2 I \subseteq I$ and $b_1 I \subseteq I$ from the inductive hypothesis. $\qquad\square$

Example 4.20. Let a and b have degree 1. The right ideal generated by $a + b$ in $\Theta(a, b)$ is not a two-sided ideal, as it does not contain $a^2 + ab$.

Definition 4.21. *Let A be a Θ-algebra and b_1, b_2 elements of $\mathcal{B}(A)$. The greatest common denominator $\gcd(b_1, b_2) \in \mathcal{B}$ and the least common multiple $\mathrm{lcm}(b_1, b_2) \in \mathcal{B}$ are defined as follows:*

$$\gcd(b_1, b_2) := gcd \text{ of } b_1, b_2 \text{ in } P(A)$$
$$\mathrm{lcm}(b_1, b_2) := lcm \text{ of } b_1, b_2 \text{ in } P(A)$$

These definitions make sense, as we shall now see:

Lemma 4.22. *Let A be a Θ-algebra and b_1, b_2 elements of $\mathcal{B}$.*

1. *If there are $a, f_1, f_2 \in A$ with $af_1 = b_1$ and $af_2 = b_2$, then there is exactly one $f \in A$ with $af = \gcd(b_1, b_2)$.*
2. *If there are $a, f_1, f_2 \in A$ with $b_1 f_1 = a$ and $b_1 f_2 = a$, then there is exactly one $f \in A$ with $\mathrm{lcm}(b_1, b_2)f = a$.*

Proof. In both parts the uniqueness of f follows from Lemma 4.13. In the first part it follows from the same lemma that a, f_1, f_2 are terms[2]. So there is an $f' \in A$ such that $a *_P f' = \gcd(b_1, b_2)$, as $P(A)$ is a polynomial algebra. But $af = \pm a *_P f'$.

Lemma 4.13 also implies that it suffices to prove the second part for terms a, f_1 and f_2. Again, we use the corresponding result for $P(A)$. □

Lemma 4.23. *Let A be a Θ-algebra and $\leq$ a useable ordering on the k-basis $\mathcal{B}(A)$. The following statements are equivalent:*

1. *$\leq$ is a well-ordering.*
2. *1 is the smallest element of $\mathcal{B}$.*

Proof. Suppose there is a $b \in \mathcal{B}$ with $b < 1$. Write b^{*n} for the nth $*_P$-power of b. So $b^{*n} = \pm b^n$ and $b^{*3} = b *_P b *_P b$. Then $1 > b > b^{*2} > \cdots > b^{*n} > b^{*n+1} > \cdots$.

Now suppose that 1 is the smallest element of $\mathcal{B}$ and let b_n be a decreasing sequence in $\mathcal{B}$. Let I be the monomial ideal in $P(A)$ generated by the b_n. As $P(A)$ is noetherian there is an $n_0 \geq 1$ such that I is generated by $b_1, \ldots, b_{n_0}$. So for each $n > n_0$ there are $m = m(n) \in \{1, \ldots, n_0\}$ and $\bar{b} = \bar{b}(n) \in \mathcal{B}$, satisfying $b_n = b_m *_P \bar{b}$. But then

$$b_m = b_m *_P 1 \leq b_m *_P \bar{b} = b_n \leq b_{n_0} \leq b_m,$$

and so $b_n = b_{n_0}$ for all $n \geq n_0$. □

---

[2] A term is a scalar multiple of a monomial, and a monomial is an element of $\mathcal{B}$.

4.2 Gröbner bases for right ideals

We now know the structure of Θ-agebras well enough to set up Gröbner bases in this context. These are modelled on Gröbner bases for ideals in (commutative) polynomial algebras, as described in Chapter 15 of Eisenbud's book [31].

4.2.1 Gröbner bases and the Division Algorithm

Definition 4.24. Let I be a right ideal in the Θ-algebra A and $f_S = (f_s)_{s \in S}$ a family of nonzero elements of I. Assume a useable ordering on $\mathcal{B}(A)$ has been chosen.

1. $LM(f_S)$ denotes the family $\{LM(f_s) \mid s \in S\}$.
2. Denote by $\langle LM(f_S) \rangle$ the ideal generated by $LM(f_S)$ in the free commutative monoid $(\mathcal{B}, *_P)$. Hence $\langle LM(f_S) \rangle$ is a k-basis for the monomial ideal $LM(f_S)A$.
3. f_S is a Gröbner basis for the right ideal I if

$$\langle LM(f_S) \rangle = LM(I \setminus \{0\}).$$

4. f_S is a minimal Gröbner basis for the right ideal I if $LM(f_S)$ is a minimal generating set for the ideal $LM(I \setminus \{0\})$ in $(\mathcal{B}, *_P)$.

Remark 4.25. Consequently minimal Gröbner bases are always finite.

Hypothesis 4.26. Let A be a Θ-algebra, I a homogeneous right ideal in A, $f_S = (f_s)_{s \in S}$ a family of homogeneous nonzero elements of I and $\leq$ a useable ordering on $\mathcal{B}(A)$.

Proposition-Definition 4.27. (cf. [31, p. 334])
Assuming Hypothesis 4.26 one has:

1. For each $f \in A$ there is an $f' \in A$, an $n \geq 0$ and triples (s_i, b_i, λ_i) for $1 \leq i \leq n$ satisfying the following conditions:
   a) $s_i \in S$, $b_i \in \mathcal{B}$ and $\lambda_i \in k^\times$ for cach i.
   b) $f = f' + \sum_{i=1}^{n} \lambda_i f_{s_i} b_i$.
   c) If $f' \neq 0$ then $LM(f') \notin \langle LM(f_S) \rangle$.
   d) $LM(f_{s_i} b_i) > LM(f_{s_j} b_j)$ for $i < j$.
   e) If $f' \neq 0$ and $n \geq 1$ then $LM(f_{s_n} b_n) > LM(f')$.
   We shall call such an f' a reduced form of f over $(\leq, f_S)$. The following properties also hold:
   f) If $n \geq 1$ then f is nonzero and $LT(f) = \lambda_1 LT(f_{s_1} b_1)$.
   g) If $s_1, \ldots, s_r$ are known for some $r \leq n$ then the b_i and the λ_i are uniquely determined for $i \leq r$. If n und $s_1, \ldots, s_n$ are known then f' is uniquely determined.

2. *For each $f \in A$ there is an $f' \in A$, an $n \geq 0$ and triples (s_i, b_i, λ_i) for $1 \leq i \leq n$ which satisfy the above conditions (a), (b) and (d) and also*
   (c') $\operatorname{supp}(f') \cap \langle LT(f_S) \rangle = \emptyset$.
   We shall call f' a completely reduced form of f over $(\leq, f_S)$. Property (g) above also holds here.

Proof. Part 1: If $f = 0$ or $LM(f) \notin \langle LM(f_S) \rangle$ then set $n = 0$ and $f' = f$. Otherwise there is a triple (s_1, b_1, λ_1) such that $f_1 := f - \lambda_1 f_{s_1} b_1$ satisfies either $f_1 = 0$ or $LM(f_1) < LM(f)$. If there are $s \neq s' \in S$ such that $LT(f)$ is divisble by $LT(f_s)$ and by $LT(f_{s'})$, then there is more than one possibility for s_1. But one s_1 has been chosen, b_1, λ_1 and f_1 are uniquely determined.

Iterating we obtain a sequence $f_0 = f, f_1, f_2, \ldots$ for which the sequence $LM(f_i)$ is strictly decreasing. This sequence (f_i) is finite, because every monomial $LM(f_i)$ lies in the finite set

$$\mu(f) = \{b \in \mathcal{B} \mid \exists b' \in \operatorname{supp}(f) \text{ with } |b| = |b'|\}.$$

Part 2: First calculate a reduced form g of f. If the intersection $\operatorname{supp}(g) \cap \langle LM(f_S) \rangle$ is not empty, set $h := g - LT(g)$. Then h is nonzero and $LM(h) < LM(f)$. We may assume that h has a completely reduced form h', for as in Part 1 the induction is well-founded. Then $LT(g) + h'$ is a completely reduced form of f. $\qquad\square$

Remark 4.28. Hypothesis 4.26 therefore implies that every Gröbner basis for the right ideal I is also a generating set for I. For the only possible reduced form for an element of I is 0.

4.2.2 The Buchberger Algorithm

The Buchberger Criterion for being a Gröbner basis will now be presented, and then the Buchberger Algorithm for constructing a Gröbner basis. On the way we shall need some more definitions.

Definition 4.29. *Assume Hypothesis 4.26. Let s, t be elements of the index set S. Lemma 4.13 implies that there are unique terms τ_s, τ_t satisfying*

$$LT(f_s)\tau_s = \operatorname{lcm}(LM(f_s), LM(f_t)) = LT(f_t)\tau_t.$$

The S-polynomial $\sigma_{st} = \sigma(f_s, f_t)$ is defined by

$$\sigma_{st} := f_s\tau_s - f_t\tau_t.$$

It follows that $b < \operatorname{lcm}(LM(f_s), LM(f_t))$ for every $b \in \operatorname{supp}(\sigma_{st})$.

Definition 4.30. *Assume Hypothesis 4.26.*

1. *A homogeneous element $f \in A$ is called weakly reducible if either $f = 0$ or there are $s_1, \ldots, s_r \in S$ and homogeneous elements $g_1, \ldots, g_r \in A \setminus \{0\}$ satisfying the following conditions:*

a) $f = \sum_{i=1}^{r} f_{s_i} g_i$.

b) $LM(f_{s_i} g_i) \leq LM(f)$ for every i.

Note in particular that f is weakly reducible if it has 0 as a reduced form.

2. Let s, t be elements of S. The S-polynomial σ_{st} is called weakly resolvable if there are $s_1, \ldots, s_r \in S$ and homogeneous elements $g_1, \ldots, g_r \in A \setminus \{0\}$ satisfying:

a) $\sigma_{st} = \sum_{i=1}^{r} f_{s_i} g_i$.

b) $LM(f_{s_i} g_i) < \text{lcm}(LM(f_s), LM(f_t))$ for every i.

So if σ_{st} is weakly reducible then it is weakly resolvable, but the reverse implication may well be false.

Recall that the generating set of a Θ-algebra is part of its structure.

Theorem 4.31 (Buchberger Criterion). (cf. [31, p. 336])

Let I be a right ideal in the Θ-Algebra A and f_S a family of elements of I, as in Hypothesis 4.26. Assume further that f_S generates I as a right ideal. Then the following statements are equivalent:

1. f_S is a Gröbner basis for the right ideal I.
2. Every reduced form of σ_{st} is zero for all $s, t \in S$.
3. σ_{st} has zero as a reduced form for all $s, t \in S$.
4. σ_{st} is weakly resolvable for all $s, t \in S$.

If moreover z^2 is weakly reducible for every odd-dimensional generator z of A, then each of the following statements is equivalent to the first four statements.

5. Statement 2 holds for all s, t with $\gcd(LM(f_s), LM(f_t)) \neq 1$.
6. Statement 3 holds for all s, t with $\gcd(LM(f_s), LM(f_t)) \neq 1$.
7. Statement 4 holds for all s, t with $\gcd(LM(f_s), LM(f_t)) \neq 1$.

We shall see in Example 4.33 that Statements 1. and 7. are not equivalent without the condition on the z^2.

Proof. Each statement follows from 1. and implies 7. Each of the first four statements also implies 4. The proof that 4. implies 1. is directly analogous to the proof of [31, Theorem 15.8].

It remains to show that 7. implies 4. Denote by $y_1, \ldots, y_N$ the odd-dimensional generators of A. For $s \in S$ write $b_s = LM(f_s)$, $\tau_s = LT(f_s)$ and $h_s = \tau_s - f_s$. Hence $f_s = \tau_s - h_s$. Suppose $s, t \in S$ satisfy $\gcd(b_s, b_t) = 1$. Define $\varepsilon \in \{+1, -1\}$ by $b_s b_t = \varepsilon b_t b_s$. Then σ_{st} is a scalar multiple of $f_s \tau_t - \varepsilon f_t \tau_s$. Showing that σ_{st} is weakly resolvable means showing that $f_s \tau_t - \varepsilon f_t \tau_s = \sum_{i=1}^{m} f_{s_i} g_i$ for suitable $s_1, \ldots, s_m \in S$ and $g_i \in A \setminus \{0\}$ satisfying $LM(f_{s_i} g_i) < b_s *_P b_t$ for each i. But

$$f_s \tau_t - \varepsilon f_t \tau_s = \varepsilon h_t \tau_s - h_s \tau_t$$
$$= \varepsilon h_t f_s - h_s f_t + (\varepsilon h_t h_s - h_s h_t)$$
$$= f_s h_t - \varepsilon f_t h_s - \Phi$$

where Φ is defined by $\Phi := (f_s h_t + h_s f_t - h_s h_t) - \varepsilon(h_t f_s + f_t h_s + h_t h_s)$. By assumption f_s, f_t are homogeneous. Since b_s, b_t have no common factor, $\varepsilon = (-1)^{|f_s| \cdot |f_t|}$. Now, Lemma 4.19 tells us that there are $\phi_1, \ldots, \phi_N$ in A that satisfy $\Phi = \sum_{j=1}^{N} j_j^2 \phi_j$ as well as for each j the condition $\operatorname{supp}(y_j^2 \phi_j) \subseteq \operatorname{supp}(\Phi)$. So if each y_j^2 is weakly reducible we obtain the required expression for $f_s \tau_t - \varepsilon f_t \tau_s$.   □

Theorem 4.32 (Buchberger Algorithm). (cf. [31, p. 337])
Let A be the Θ-algebra $\Theta(x_1, \ldots, x_M, y_1, \ldots, y_N)$, where each x_i is even- and each y_j is odd-dimensional. Let $\leq$ be a useable ordering on $\mathcal{B}(A)$ and I a homogeneous right ideal in A. Let $f_S = (f_s)_{s \in S}$ be a finite family of homogeneous nonzero elements of A which generates I.

Assume further that each y_j^2 is contained in I and weakly reducible over f_S.

If at least one step of the following two types is possible, then carry one out to obtain a new family f_U.

Type 1 step *Pick an unordered pair $\{s, t\} \subseteq S$ and calculate a reduced form f' over f_S of its S-polynomial σ_{st}. If $f' = 0$ then set $U := S$, otherwise set $U := S \amalg \{u_0\}$ and $f_{u_0} := f'$. Set $T := S$ in both cases.*
Qualification: Only use pairs which satisfy $\gcd(LM(f_s), LM(f_t)) \neq 1$ and for which this step has not been carried out before.

Type 2 step *Pick $s \in S$, set $T := S \setminus \{s\}$ and calculate a reduced form f' of f_s over f_T. Set $U := T$ if $f' = 0$, otherwise set $U := T \amalg \{u_0\}$ and $f_{u_0} := f'$.*
Qualification: $LM(f_s)$ must lie $\langle LM(f_T) \rangle$. That is, there is a $t \in T$ and a term τ satisfying $LT(f_s) = LT(f_t)\tau$.

We then obtain a sequence $f_{S(0)}, f_{S(1)}, \ldots$ by setting $S(0) = S$ and $f_{S(n+1)}$ to be the f_U for $f_{S(n)}$. This sequence has finite length and the last term $f_{S(n)}$ is a minimal Gröbner basis for I.

Proof. The monomial ideal $\langle LT(f_U) \rangle$ contains $\langle LT(f_S) \rangle$, and they are equal if and only if $T = U$. As monomial ideals are also ideals in the polynomial algebra $P(A)$ they satisfy the ascending chain condition. So after finitely many steps one has $T = U$. But then the sequence stops after finitely many steps: if $T = U$ for a Type 2 step then U is a proper subset of S, and if $T = U$ for a Type 1 step then $U = S$ and there are only finitely many pairs s, t to work through.

Clearly f_U and f_S generate the same right ideal in A. If $f \in A$ is weakly reducible over f_S, then it is weakly reducible over f_U too. Similarly for $t, t' \in T$: $\sigma_{tt'}$ is weakly reducible over f_U if it is over f_S. For the last term $f_{S(n)}$ of the sequence no Type 1 steps are possible, and so Statement 7. of Theorem 4.31 is satisfied. The assumption on the odd-dimensional generators ensures then that $f_{S(n)}$ is a Gröbner basis for I. Since no Type 2 steps are possible, this Gröbner basis is minimal.   □

Example 4.33. Let A be $\Theta(a, b, c, d, x, y)$, where the generator degrees are $|x| = |y| = 2$, $|a| = |b| = 1$ and $|c| = |d| = 3$. Assume the ground field k has characteristic greater than 2. Define $f_i \in A$ for $1 \le i \le 12$ as follows:

$$f_1 := a^2 \qquad f_2 := b^2 \qquad f_3 := ax - c - d \qquad f_4 := by - c + d$$

and

$$f_5 := ac + ad \qquad f_6 := bc - bd \qquad f_7 := c^2 + d^2 \qquad f_8 := abd$$
$$f_9 := bd^2 \qquad f_{10} := ad^2 \qquad f_{11} := cd^2 - d^3 \qquad f_{12} := d^3$$

Set $S := \{1, 2, 3, 4\}$ and $R := \{1, \ldots, 12\}$, and let I be the right ideal generated by f_S. Applying the Buchberger Algorithm to f_S produces f_R. The equations

$$f_3^2 - f_1 x^2 = c^2 + d^2 \quad \text{and} \quad f_3 f_4 + f_4 f_3 = 2(c^2 - d^2)$$

imply that c^2 and d^2 lie in I. This means that f_R cannot possibly be a Gröbner basis for I, for $\langle LM(f_R) \rangle$ does not contain d^2. But f_R does satisfy Statement 7. of Theorem 4.31. We deduce that c^2 and d^2 are not weakly reducible over f_R, and therefore not over f_S either.

Conclusion: At the start of the Buchberger Algorithm it is essential to ensure that y^2 is weakly reducible for each odd-dimensional generator y. The simplest way to ensure this is to append the y^2 to the family f_S.

4.3 The kernel of an algebra homomorphism

At one point in Carlson's Completeness Criterion one has to compute the kernel of the restriction map from $\mathrm{H}^*(G)$ to $\mathrm{H}^*(H)$ for each maximal subgroup $H \le G$. So we need to be able to determine the kernel of a map of graded-commutative algebras.

Let $A = \Theta(z_1, \ldots, z_n)$ and $B = \Theta(w_1, \ldots, w_m)$ be Θ-algebras, let $I \subseteq A$ and $J \subseteq B$ be two-sided ideals and let $\phi : A/I \to B/J$ be a map of graded algebras. It is well known that in the commutative case the kernel of ϕ can be determined by elimination. The first step is to lift ϕ to a map $\Phi : A \to B$ of algebras.

Great care has to be taken when applying elimination methods to the case where A and B are Θ-algebras. Firstly, it is not always possible to lift to a map of graded algebras $\Phi : A \to B$ (cf. Example 4.34). Secondly, Example 4.35 shows that the evaluation map $A \otimes_k B \to B/J$ is not in general a map of graded algebras. But if the algebras A/I and B/J are both graded commutative, then the kernel can be computed.

Example 4.34. Let $A = B = \Theta(y_1, y_2)$ with $|y_i| = 1$, and let $I = J = (y_1^2, y_2^2)$. Define a map of graded algebras $\phi : A/I \to B/J$ by $y_1 + I \mapsto y_1 + y_2 + J$ and $y_2 + I \mapsto y_2 + J$. The only possibility for a lift Φ is Φ is $y_1 \mapsto y_1 + y_2$ and $y_2 \mapsto y_2$. But then Φ does not respect the relation $y_2 y_1 = -y_1 y_2$.

For $\phi: A/I \to B/J$ as above, let C be the Θ-algebra $\Theta(w_1, \ldots, w_m, z_1, \ldots, z_n)$ which contains both A and B as subalgebras. Let $\Gamma \subseteq C$ be the two-sided ideal generated by

$$\{b - a \mid a \in A, \, b \in B \text{ and } \phi(a + I) = b + J\}.$$

We must calculate the two-sided ideal $K := \{a \in A \mid a + I \in \mathrm{Ker}(\phi)\}$ in A.

Example 4.35. Let $A = B = \Theta(y)$ with $|y| = 1$ and let $\phi: A \to B$ be the identity map. Then C is the Θ-algebra $\Theta(y_B, y_A)$. There is no map $C \to B$ of graded algebras sending both y_A and y_B to y, as it would be impossible to preserve the relation $y_B y_A = -y_A y_B$.

Lemma 4.36. *If B/J is graded commutative then $A \cap \Gamma = K$.*

Proof. Clearly $K \subseteq A \cap \Gamma$. Now let $\eta: C \to B/J$ be the map of graded algebras which sends w_j to $w_j + J$ and z_i to $\phi(z_i + I)$. This is indeed a map of graded algebras, because the graded commutativity of B/J means that the relation $z_i \cdot w_j = (-1)^{|z_i| \cdot |w_j|} w_j \cdot z_i$ is preserved. It is immediate from the definitions that $\Gamma \subseteq \mathrm{Ker}(\eta)$. Equally it is clear that $A \cap \mathrm{Ker}(\eta)$ lies in K. $\square$

The elimination ordering Lemma 4.36 means that the kernel of a map of graded commutative algebras can be found by elimination. Elimination is a special case of the Buchberger Algorithm which uses an elimination ordering.

Definition 4.37. *Let $\leq$ be a useable ordering on $\mathcal{B}(C)$. Each $b \in \mathcal{B}(C)$ is assigned an elimination dimension $\mathrm{edim}(b)$ by setting $\mathrm{edim}(z_i) = 0$ and $\mathrm{edim}(w_j) = |w_j|$. The elimination ordering $\leq_{\mathrm{elim}}$ on $\mathcal{B}(C)$ is then defined for all $b, c \in \mathcal{B}(C)$ as follows:*

$$b \leq_{\mathrm{elim}} c :\Longleftrightarrow |b| < |c|; \text{ or}$$
$$|b| = |c| \text{ and } \mathrm{edim}(b) < \mathrm{edim}(c); \text{ or}$$
$$|b| = |c|, \mathrm{edim}(b) = \mathrm{edim}(c) \text{ and } b \leq c.$$

Corollary 4.38. *Suppose we are given Θ-algebras $A = \Theta(z_1, \ldots, z_n)$ and $B = \Theta(w_1, \ldots, w_m)$, together with two-sided homogeneous ideals $I \subseteq A$ and $J \subseteq B$ such that B/J is a graded commutative algebra. Let $\phi: A/I \to B/J$ be a map of graded algebras and let $\leq$ be a useable ordering on $\mathcal{B}(C)$, where we set $C := \Theta(w_1, \ldots, w_m, z_1, \ldots, z_n)$. Let $\Gamma \subseteq C$ be the two-sided ideal generated by the set*

$$\{b - a \mid a \in A, \, b \in B \text{ and } \phi(a + I) = b + J\}$$

Suppose $f_S := (f_s)_{s \in S}$ is a family of homogeneous elements of C which form a minimal Gröbner basis for the right ideal Γ with respect to the elimination ordering $\leq_{\mathrm{elim}}$ on $\mathcal{B}(C)$. Set $T := \{s \in S \mid LM_{\leq_{\mathrm{elim}}}(f_s) \in \mathcal{B}(A)\}$. Then f_T is a minimal Gröbner basis for the homogeneous right ideal

$$K := \{a \in A \mid a + I \in \mathrm{Ker}(\phi)\}.$$

Proof. The elimination ordering is constructed to ensure that $LM_{\leq_{\mathrm{elim}}}(f_s)$ lies in $\mathcal{B}(A)$ if and only if $f_s \in A$.   □

Example 4.39. Choose the ordering $\leq_{\mathrm{grlex}}$ of Example 4.10 on $\mathcal{B}(C)$. Let $\leq_{\mathrm{elim}}$ be the corresponding elimination ordering. The following orderings on $\mathcal{B}(A)$ coincide:

1. The ordering $\leq_{\mathrm{grlex}}$ on $\mathcal{B}(A)$;
2. The restriction to $\mathcal{B}(A)$ of the ordering $\leq_{\mathrm{grlex}}$ on $\mathcal{B}(C)$;
3. The restriction to $\mathcal{B}(A)$ of the ordering $\leq_{\mathrm{elim}}$ on $\mathcal{B}(C)$.

Hence the family f_T in Corollary 4.38 is a minimal Gröbner basis for K with respect to the usual graded-lexicographical ordering $\leq_{\mathrm{grlex}}$ on $\mathcal{B}(A)$.

Example 4.40. The ordering $\leq_{\mathrm{coho}}$ is defined in Definition 5.9. If $\mathcal{B}(A)$ and $\mathcal{B}(B)$ are ordered with this ordering, then one can order $\mathcal{B}(C)$ with this ordering: for the y-degree and the r-dimension of each generator of C are known. Now let $\leq_{\mathrm{elim}}$ be the corresponding elimination ordering on $\mathcal{B}(C)$. Then the following orderings on $\mathcal{B}(A)$ coincide:

1. The ordering $\leq_{\mathrm{coho}}$ on $\mathcal{B}(A)$;
2. The restriction to $\mathcal{B}(A)$ of the ordering $\leq_{\mathrm{coho}}$ on $\mathcal{B}(C)$;
3. The restriction to $\mathcal{B}(A)$ of the ordering $\leq_{\mathrm{elim}}$ on $\mathcal{B}(C)$.

So the family f_T of Corollary 4.38 is a minimal Gröbner basis for K with respect to the ordering $\leq_{\mathrm{coho}}$ on $\mathcal{B}(A)$. This is the ordering used in the package Diag.

Remark 4.41. In these two cases (Examples 4.39 and 4.40) one can therefore use the Buchberger Algorithm (Theorem 4.32) to construct a minimal Gröbner basis for K, provided that the algebra A/I is also graded commutative. For this we have to construct a finite family f_S of homogeneous elements of C which generates Γ as a right ideal. An f_S of this kind can be obtained as the union of three families:

− A minimal Gröbner basis for I;
− A minimal Gröbner basis for J;
− $\{h_i - z_i \mid 1 \leq i \leq n\}$. Here h_i is the unique completely reduced element of the coset $\phi(z_i + I) \in B/J$.

This f_S contains minimal Gröbner bases for I and J, and so y^2 is weakly reducible over f_S for each odd-dimensional $y \in \{w_1, \ldots, w_m, z_1, \ldots, z_n\}$. So the right ideal generated by f_S is two-sided. Moreover $(b_1 - a_1)b_2 + a_1(b_2 - a_2) = b_1 b_2 - a_1 a_2$ for $a_1, a_2 \in A$ and $b_1, b_2 \in B$ with $b_i + J = \phi(a_i + I)$. Since $b_1 b_2 + J = \phi(a_1 a_2 + I)$, the ideal generated by f_S is indeed Γ. The hypotheses of Theorem 4.32 are satisfied.

4.4 Intersections and Annihilators: Gröbner bases for modules

We shall now generalise Gröbner bases for right ideals in Θ-algebras to Gröbner bases for right modules over Θ-algebras. Again, this is modelled on the commutative case: indeed Eisenbud [31] defines Gröbner bases for modules straight away: Gröbner bases for ideals are then merely a very important special case.

Gröbner bases for modules have many uses. The current version of the package Diag uses them for two purposes: to compute the intersection of two ideals and to determine the annihilator of an element.

This section contains nothing really new: the underlying methods are very well known, and there are no problems adapting them to Θ-algebras.

Definition 4.42. *Let A be a Θ-algebra, $\leq$ a useable ordering on $\mathcal{B}(A)$ and $F := \bigoplus_{i=1}^{m} e_i A$ the free right A-module on generators $e_1, \ldots, e_m$.*

1. *Denote by $\mathcal{B}(F)$ the following k-basis of F:*

   $$\mathcal{B}(F) := \{e_i b \mid 1 \leq i \leq s, \quad b \in \mathcal{B}(A)\}.$$

2. *The operation $*_P \colon \mathcal{B}(F) \times \mathcal{B}(A) \to \mathcal{B}(F)$ defined by $(e_i b) *_P b' := e_i(b *_P b')$ makes $\mathcal{B}(F)$ a k-basis of the free module $P(F)$ on the e_i over the associated polynomial algebra $P(A)$. Moreover*

   $$(e_i b)b' = e_i(bb') = \pm e_i(b *_P b').$$

   The word "monomial" will also be used for elements of $\mathcal{B}(F)$.

3. *An ordering $\leq$ on $\mathcal{B}(F)$ is useable if for all $e_i b, e_j c \in \mathcal{B}(F)$ and $b', c' \in \mathcal{B}(A)$*

   If $e_i b \leq e_j c$ and $b' \leq c'$ then $(e_i b) *_P b' \leq (e_j c) *_P c'$.

4. *Given a useable ordering on $\mathcal{B}(F)$, the support $supp(f) \subseteq \mathcal{B}(F)$ and the leading monomial $LM(f) \in \mathcal{B}(F)$ of a nonzero $f \in F$ are defined in the usual way.*

5. *Let $M \subseteq F$ be a submodule and $f_S = (f_s)_{s \in S}$ a family of nonzero elements of M. Define $LM(f_S)$ and $\langle LM(f_S) \rangle$ as in Definition 4.24, and also the concepts Gröbner basis and minimal Gröbner basis for M. In particular, $\langle LM(f_S) \rangle$ is the submodule of the free $(\mathcal{B}(A), *_P)$-module $\mathcal{B}(F)$ generated by $LM(f_S)$.*

6. *Assigning each generator e_i of F a degree $|e_i| \in \mathbb{Z}$ means that every $e_i b \in \mathcal{B}(F)$ receives a degree. Set $\mathcal{B}(F)_n := \{e_i b \in \mathcal{B}(F) \mid |e_i b| = n\}$ and $F_n := \{f \in F \mid supp(f) \subseteq \mathcal{B}(F)_n\}$. Then $F = \bigoplus_{n \in \mathbb{Z}} F_n$. Moreover $\mathcal{B}(F)_n$ is finite and $F_n = \{0\}$ for n sufficiently large and negative. If $f \in F_n$ then f is homogeneous of degree n.*

Example 4.43. Given a useable ordering on $\mathcal{B}(A)$ and an ordering on the free generators $e_1, \ldots, e_m$ of F we obain a useable ordering on $\mathcal{B}(F)$ by setting

$$e_i b \leq e_j c \iff e_i < e_j \quad \text{or} \quad e_i = e_j \text{ and } b \leq c.$$

Example 4.44. (The elimination ordering)
Let F' and F'' be free right A-modules, and assume useable orderings on $\mathcal{B}(A)$, $\mathcal{B}(F')$ and $\mathcal{B}(F'')$ have been chosen. Setting $F := F' \oplus F''$ we have $\mathcal{B}(F) = \mathcal{B}(F') \amalg \mathcal{B}(F'')$. We extend the orderings on $\mathcal{B}(F')$ and $\mathcal{B}(F'')$ to an ordering on $\mathcal{B}(F)$ by setting $b'' < b'$ for all $b' \in \mathcal{B}(F')$ and $b'' \in \mathcal{B}(F'')$. For $f \in F \setminus \{0\}$ one then has: $LM(f) \in \mathcal{B}(F'')$ if and only if $f \in F''$.

Now let $M \subseteq F$ be a submodule and f_S a Gröbner basis for M. Setting $T := \{s \in S \mid f_s \in F''\}$ we obtain a Gröbner basis f_T for $M \cap F''$. If f_S is minimal for M then f_T is minimal for $M \cap F''$.

Example 4.45. (The intersection of two right ideals)
Let I and J be right ideals in A. Set $F' := A$, $F'' := A$ and $F := F' \oplus F''$. Order $\mathcal{B}(F)$ with the elimination ordering. Define a submodule M of F by

$$M := \{(i + j, i) \in F \mid i \in I \text{ and } j \in J\}.$$

Then $M \cap F''$ is the set $\{a \in A \mid (0, a) \in M\}$, that is $I \cap J$. If f_S is a Gröbner basis for M, set

$$T := \{s \in S \mid \text{There is a } g_s \in A \text{ with } f_s = (0, g_s)\}.$$

This defines a family g_T which is a Gröbner basis for the right ideal $I \cap J \subseteq A$.

Example 4.46. (The annihilator of an element)
Given a two-sided homogeneous ideal $I \subseteq A$ and a homogeneous element $\zeta \in A$ we need to be able to calculate the annihilator $\mathrm{Ann}_{A/I}(\zeta)$. We shall do this by calculating the right ideal $J \subseteq A$ defined by $J := \{a \in A \mid \zeta a \in I\}$. Since I is two-sided, $I \subseteq J$ and one has $J/I = \mathrm{Ann}_{A/I}(\zeta)$.

Again set $F := A \oplus A$ with the appropriate elimination ordering and consider the submodule $M := \{(i + \zeta a, a) \mid i \in I \text{ and } a \in A\}$ of F. Then $M \cap F'' = \{a \in A \mid (0, a) \in M\}$ is the desired ideal J. If f_S is a Gröbner basis for M then set

$$T := \{s \in S \mid \text{There is a } g_s \in A \text{ mit } f_s = (0, g_s)\}.$$

This defines a family g_T which is a Gröbner basis for the right ideal J.

Remark 4.47. The Buchberger Algorithm for right ideals in Θ-algebras assumed that the ideal was homogeneous. This is because there are basically two ways to ensure that the Division Algorithm in Proposition-Definition 4.27 and the Buchberger Algorithm 4.32 stop: either one requires that the useable ordering on $\mathcal{B}(A)$ is a well-ordering or one restricts to homogeneous ideals and uses the fact that each $\mathcal{B}(A)_n$ is finite.

We shall also only establish the Buchberger Algorithm for modules in the homogeneous case. Note however that Definition 4.42 allows the degrees of the module generators to be nonzero and to differ from each order. This is sensible and indeed necessary, as Example 4.46 shows. Here the generators of F are $e_1 := (1,0)$ and $e_2 := (0,1)$. If we set $|e_1| := -|\zeta|$ and $|e_2| := 0$ then $(\zeta a, a)$ is homogeneous of degree n for all $a \in A_n$. Hence the submodule M is homogeneous.

Remark 4.48. Let I be a two-sided ideal in the Θ-algebra A. More generally, Gröbner bases and an elimination ordering can be used to determine the kernel of an arbitrary map of free right A/I-modules.

Remark 4.49. Let I and J be homogeneous right ideals in the Θ-algebra A, and assume Gröbner bases f_S for I and g_T for J are known. How should the computer decide whether $I \subseteq J$? Answer: By computing a reduced form f_s' over g_T of each f_s. The Buchberger Criterion (Theorem 4.31) says that $I \subseteq J$ if and only if $f_s' = 0$ for all $s \in S$. This method could equally be applied to any generating set of I.

Hypothesis 4.50. Let A be a Θ-algebra and $F := \bigoplus_{i=1}^m e_i A$ the free right A-module on generators $e_1, \dots, e_m$. Suppose we are given compatible useable orderings on $\mathcal{B}(A)$ and on $\mathcal{B}(F)$. Let $M \subseteq F$ be a submodule and $f_S = (f_s)_{s \in S}$ a family of homogeneous elements of $M \setminus \{0\}$.

Definition 4.51. *Assume Hypothesis 4.50. Let s and t be elements of S. Write $LM(f_s) = e_{i_s} b_s$.*

If $i_t \neq i_s$ then there is no S-polynomial σ_{st}. If $i_t = i_s =: i$, then Lemma 4.13 says there are unique terms τ_s, τ_t satisfying

$$LT(f_s)\tau_t = e_i \mathrm{lcm}(b_s, b_t) = LT(f_t)\tau_s .$$

The S-polynomial $\sigma_{st} = \sigma(f_s, f_t)$ is defined by $\sigma_{st} := f_s \tau_t - f_t \tau_s$. Hence $c < e_i \mathrm{lcm}(b_s, b_t)$ for every $c \in supp(\sigma_{st})$. The definition of "weakly resolvable" for S-polynomials for modules is analogous to Definition 4.30.

Theorem 4.52 (Buchberger Criterion for modules). (cf. [31, p. 336]) *Assume Hypothesis 4.50. Moreover let M be the submodule of f_S generated by F. Denote by $\mathcal{C}$ the set of all critical pairs*

$$\mathcal{C} := \{(s,t) \in S^2 \mid \text{There is an S-polynomial } \sigma_{st}\} .$$

The following statements are equivalent:

1. *f_S is a Gröbner basis for the module M.*
2. *Every reduced form of σ_{st} is zero for each $(s,t) \in \mathcal{C}$.*
3. *For each $(s,t) \in \mathcal{C}$, zero is a reduced form of $f = \sigma_{st}$.*
4. *σ_{st} is weakly resolvable for each $(s,t) \in \mathcal{C}$.*

Proof. Each statement follows from 1. and implies 4. The proof that 4. implies 1. is directly analogous to the proof of [31, Theorem 15.8]. $\qquad\square$

Example 4.53. Let A be the Θ-algebra $\Theta(x, y)$ with $|x| = |y| = 2$. So A is the polynomial algebra $k[x, y]$. Let F be the free right A-module $e_1 A \oplus e_2 A$ and let $M \subseteq F$ be the submodule generated by $f_1 := e_1 x$ and $f_2 := e_1 y - e_2 x$. Then $f_1 y - f_2 x = e_2 x^2$. This is an example of why there are only four equivalent statements in Theorem 4.52, and not seven as in Theorem 4.31.

Theorem 4.54 (Buchberger Algorithm for modules). (cf. [31, p. 337])
Assume Hypothesis 4.50. Moreover assume f_S generates the submodule M.

If at least one step of the following two types is possible, then carry one out to obtain a new family f_U.

Type 1 step *Pick an unordered pair $s, t \in S$ with $s \neq t$ and calculate a reduced form f' over f_S of its S-polynomial σ_{st}. If $f' = 0$ then set $U := S$. Otherwise set $U := S \amalg \{u_0\}$ and $f_{u_0} := f'$. Set $T := S$ in both cases.*
Qualifications: The S-polynomial σ_{st} exists. This step has not yet been carried out on this pair.

Type 2 step *Pick an $s \in S$, set $T := S \setminus \{s\}$ and calculate a reduced form f' of f_s over f_T. If $f' = 0$ then set $U := T$. Otherwise set $U := T \amalg \{u_0\}$ and $f_{u_0} := f'$.*
Qualification: $LM(f_s)$ lies in $\langle LM(f_T) \rangle$. That is, there is a $t \in T$ and a term τ satisfying $LT(f_s) = LT(f_t)\tau$.

We then obtain a sequence $f_{S(0)}, f_{S(1)}, \ldots$ of families by setting $S(0) = S$ and $f_{S(n+1)}$ to be the f_U for $f_{S(n)}$. This sequence has finite length and the last term $f_{S(n)}$ is a minimal Gröbner basis for M.

Proof. Analogous to the proof of Theorem 4.32. $\qquad\square$

5 The visible ring structure

Once the first N terms of the minimal resolution are known, one can determine the presentation (generators and relations) of the cohomology ring out to degree N. The method was developed by Carlson, E. Green and Schneider [21], and is recalled in Sect. 5.1. Then in Sect. 5.2 we will discuss how Gröbner bases for kG-modules can be used to carry out this method.

In order to make the relations in the cohomology ring as simple as possible, one has to take care how one chooses generators. One strategy for choosing generators is described in Sect. 5.3. One also wants the Gröbner basis for the relations ideal to be as small as possible, so this section also introduces a useable monomial ordering which was designed in tandem with the generator choice strategy.

In Sect. 5.4 we shall consider a technical question that arises when determining the relations: how best to compute a large batch of product cocycles in one go. Finally, Sect. 5.5 concerns the restriction map to the cohomology ring of a subgroup.

5.1 Basics

Let (P_n, d_n) be the minimal projective resolution of the trivial kG-module k. Then P_n is a free kG-module, whose rank we shall denote by $\beta(n)$. The nth cohomology group $\mathrm{H}^n(G)$ is the additive group $\mathrm{Hom}_{kG}(P_n, k)$. Since P_n is a free kG-module, cohomology classes can be represented as vectors in $k^{\beta(n)}$, where the ith component of such a vector is the value of the corresponding cocycle on the ith free generator.

Each such cocycle $\phi\colon P_n \to k$ can be lifted to a kG-linear map $\phi_0\colon P_n \to P_0$, which satisfies $\varepsilon \circ \phi_0 = \phi$. Of course, the map $\varepsilon\colon P_0 = kG \to k$ here is the augmentation map, which is the projective cover of k. In turn, one can lift ϕ_0 to a chain map

$$
\begin{array}{ccccccccc}
\cdots & \longrightarrow & P_{n+m} & \xrightarrow{d_{n+m}} & P_{n+m-1} & \longrightarrow & \cdots & \longrightarrow & P_n & \xrightarrow{d_n} & P_{n-1} \\
& & \phi_m \downarrow & & \phi_{m-1} \downarrow & & & & \phi_0 \downarrow & & \\
\cdots & \longrightarrow & P_m & \xrightarrow{d_m} & P_{m-1} & \longrightarrow & \cdots & \longrightarrow & P_0 & &
\end{array}
$$

Now let $\psi\colon P_m \to k$ be a degree m class. The product[1] $\psi\phi$ is the composition $P_{n+m} \overset{\phi_m}{\to} P_m \overset{\psi}{\to} k$.

One way to determine the ring structure in degrees $\leq N$ is as follows:

Algorithm 5.1.
Input: Minimal resolution $P_N \overset{d_N}{\to} P_{N-1} \to \cdots \to P_0 \overset{\varepsilon}{\to} k$.
Output: List $\mathcal{G}$ of all generators and family f_S of all relations in degrees $\leq N$.

$\mathcal{G} := \emptyset,\ S := \emptyset$.
FOR $1 \leq n \leq N$ DO
    $\mathcal{P}_n :=$ All products in degree n of elements of $\mathcal{G}$.
    Realise each element of $\mathcal{P}_n$ as a cocycle.
    $\mathcal{Z}_n :=$ subspace of $\mathrm{H}^n(G)$ spanned by these cocycles.
    Read off new relations and add to f_S.
    Extend $\mathcal{G}$ by adding on a basis for a complement of $\mathcal{Z}_n$.
END FOR

The remainder of this chapter is mainly about putting this algorithm into practice. In particular, the individual steps of the algorithm will be explained in detail in §5.2.2.

5.2 Practical considerations

5.2.1 Lifts of cocycles

Suppose we are given a map $\phi_{m-1}\colon P_{n+m-1} \to P_{m-1}$, such that the image of $\phi_{m-1}\circ d_{n+m}$ lies in $\mathrm{Im}(d_m)$. We need to find a kG-linear map $\phi_m\colon P_{n+m} \to P_m$ which makes the diagram

$$
\begin{array}{ccc}
P_{n+m} & \xrightarrow{\ d_{n+m}\ } & P_{n+m-1} \\
{\scriptstyle \phi_m}\downarrow & & \downarrow{\scriptstyle \phi_{m-1}} \\
P_m & \xrightarrow{\ d_m\ } & P_{m-1}
\end{array}
$$

commute. Let $e_1,\ldots,e_r$ and $\eta_1,\ldots,\eta_s$ be the free generators of the free kG-modules P_{n+m} and P_{n+m-1} respectively. The computer constructs a lift ϕ_m by setting

$$\phi_m(e_i) := \text{a preimage of } \phi_{m-1}d_{n+m}(e_i) \text{ under } d_m.$$

Algorithm 3.37 is used to compute preimages, but we have to consider how to determine the composition $\phi_{m-1} \circ d_{n+m}$.

---

[1] This is the Yoneda product. Here it coincides with the cup product.

The map d_{n+m} is represented on the computer by the images $d_{n+m}(e_1)$, $\ldots$, $d_{n+m}(e_r)$. In other words, d_{n+m} is represented by its matrix over kG with respect to the bases e_i and η_j. The map ϕ_{m-1} is also represented by its matrix, and so we should compute the composition by multiplying these two matrices. But we have already seen in Sect. 1.4 how to multiply elements of kG.

Calculating a product Suppose cocycles $\phi\colon P_n \to k$ and $\psi\colon P_m \to k$ are given, and that the lift $\phi_m\colon P_{n+m} \to P_m$ is known. Then the product $\psi.\phi$ is the cocycle $\psi \circ \phi_m\colon P_{n+m} \to k$. This composition can of course be determined without multiplying matrices over kG.

Let $e_1, \ldots, e_r$ be free generators for P_{n+m} and $\zeta_1, \ldots, \zeta_t$ free generators for P_m. The map ϕ_m is known to the computer by its matrix M over kG with respect to these bases. Let $\bar{M}$ be the matrix over k obtained from M by applying the augmentation map to each entry.

The cocycle ψ is represented by a vector in k^t. Applying the matrix $\bar{M}$ to this vector yields the vector in k^r which represents the cocycle $\psi.\phi$.

5.2.2 Gröbner bases and the visible ring structure

The Algorithm 5.1 will now be described in more detail. Care must be taken with the way several steps are performed.

The computer uses Gröbner bases to be able to work with relations in the cohomology ring. The type of Gröbner bases involved depends on the prime number p.

p odd: Gröbner bases for right ideals in a Θ-algebra are used, as described in Chap. 4.

Let A be the Θ-algebra $\Theta(\mathcal{G})$ and let $\leq$ be a useable ordering on the k-basis $\mathcal{B}(A)$ for A. One difficulty here is that the generating set $\mathcal{G}$ is not completely known until the end of the algorithm. This is not a problem if we use the ordering $\leq_{\mathrm{grlex}}$ (see Example 4.10), as we kann simply define each new generator to be larger (lexicographically) than each previously known generator. Equally unproblematic is the ordering $\leq_{\mathrm{coho}}$ which we shall meet in Definition 5.9: this is the ordering that actually gets used in the package Diag.

Suppose that the computer has just found a new generator y in odd degree. As Example 4.33 demonstrates, it is very important that the Relation y^2 is immediately added to the relations in f_S.

$p = 2$: In this case ordinary commutative Gröbner bases are used. Let A be the polynomial algebra $k[\mathcal{G}]$. For the sake of uniformity, we will stick as far as possible to the notation introduced for Θ-algebras. So $\mathcal{B}(A)$ stands for the free commutative monoid on $\mathcal{G}$, a k-basis for A. The associated polynomial algebra $P(A)$ now coincides with A, and the product $*_P$ agrees with the product on A. The definition of a useable ordering on $\mathcal{B}(A)$ is then obvious, and corresponds to the definition of a monomial ordering in [31, S. 328], except that useable orders are not required to be well orderings. As we however are only interested in homogeneous ideals, this difference is irrelevant.

The following algorithm is a special case of Algorithm 5.1. It uses Gröbner basis methods and is explicit enough to be implemented on the computer. Note that Type 1 steps (calculate and reduce S-polynomial) from the Buchberger Algorithm for Θ-algebras are used, but that Type 2 steps (resolve an inclusion ambiguity) are never used.

Algorithm 5.2.
Input: Minimal resolution $P_N \xrightarrow{d_N} P_{N-1} \to \cdots \to P_0 \xrightarrow{\varepsilon} k$.
Output: List $\mathcal{G}$ of all generators in degree $\leq N$. Minimal Gröbner basis f_S for the relations in degree $\leq N$, together with a minimal generating set for these relations.

$\mathcal{G} := \emptyset$, $S := \emptyset$.
FOR $1 \leq n \leq N$ DO
    Compute and reduce all S-polynomials of degree n.
    $\mathcal{P}_n :=$ all elements of $\mathcal{B}(A)$ which are irreducible and lie in degree n.
    For each $w \in \mathcal{P}_n$ calculate the cocycle ζ_w.
    $\mathcal{Z}_n :=$ the subspace of $\mathrm{H}^n(G)$ spanned by $(\zeta_w)_{w \in \mathcal{P}_n}$.
    IF $\dim_k(\mathcal{Z}_n) < |\mathcal{P}_n|$ THEN
        Append new relations to f_S, using procedure ReadOffRelations.
    END IF
    IF $\dim_k(\mathcal{Z}_n) < \dim_k(\mathrm{H}^n(G))$ THEN
        Choose new generators and add them to $\mathcal{G}$,
        using procedure ChooseGenerators.
        IF p and n are both odd THEN
            For each new $y \in \mathcal{G}$ append the relation y^2 to f_S.
        END IF
    END IF
END FOR
Finish off applying the Buchberger Algorithm to f_S.

We have already seen how to compute the product of two cocycles. In Sect. 5.4 we shall discuss efficient methods for the mass production of product cocycles ζ_w.
    The vector space $\mathcal{Z}_n$ consists of all the decomposable classes in degree n.

Procedure 5.3 (ReadOffRelations).

$\mathcal{Q} := \emptyset$
FOR w runs through $\mathcal{P}_n$ in increasing order DO
    IF there are scalars $\lambda_v \in k$ with $\zeta_w = \sum_{v \in \mathcal{Q}} \lambda_v \zeta_v$ THEN
        Append $w - \sum_{v \in \mathcal{Q}} \lambda_v v$ to f_S.

ELSE

　　Append w to $\mathcal{Q}$.

END IF

END FOR

*Procedure 5.4 (*ChooseGenerators*).*
(The use of this particular method is justified in Sect. 5.3.)
Let $Z \leq G$ be the greatest central elementary abelian subgroup.

Calculate $\mathcal{N}_n := \mathrm{H}^n(G) \cap \sqrt{0}$.

$\mathcal{Y}_n := \mathcal{Z}_n + \mathcal{N}_n$.

IF $\dim_k(\mathcal{Y}_n) > \dim_k(\mathcal{Z}_n)$ THEN

　　Determine a basis for a complement of $\mathcal{Z}_n \cap \mathcal{N}_n$ in $\mathcal{N}_n$.

　　Append this basis to $\mathcal{G}$.

END IF

Calculate $\mathcal{I}_n := \{x \in \mathrm{H}^n(G) \mid \mathrm{Res}_Z(x) \text{ is nilpotent}\}$.

$\mathcal{X}_n := \mathcal{Y}_n + \mathcal{I}_n$

IF $\dim_k(\mathcal{X}_n) > \dim_k(\mathcal{Y}_n)$ THEN

　　Determine a basis for a complement of $\mathcal{Y}_n$ in $\mathcal{X}_n$.

　　Append this basis to $\mathcal{G}$.

END IF

IF $\dim_k(\mathrm{H}^n(G)) > \dim_k(\mathcal{X}_n)$ THEN

　　Determine a basis for a complement of $\mathcal{X}_n$ in $\mathrm{H}^n(G)$.

　　Append this basis to $\mathcal{G}$.

END IF

Lemma 5.5. *At the end of Algorithm 5.2 the induced homomorphism $A^n \to \mathrm{H}^n(G)$ is surjective for $n \leq N$. Moreover, the kernel in this degree is $A^n \cap J$, where J denotes the ideal generated by f_S.*

At no point in the algorithm is a Type 2 step (resolution of an inclusion ambiguity) performed. So no elements are ever removed from the index set S, and the value of no f_s is ever changed.

Hence there are three kinds of relations f_s:

1. *Reduced forms of S-polynomials.*
2. *Relations produced by* ReadOffRelations.
3. *Relations of the form y^2, where p and $|y|$ are both odd.*

The relations of the last two kinds constitute a minimal generating set for J.

Proof. The relation family f_S is determined degree by degree, and so the selection method in ReadOffRelations ensures that inclusion ambiguities[2] never arise.

　　Strictly speaking though, f_S is not built up degree by degree when p is odd, for the relations y^2 with $|y|$ odd are recorded too early. However, an

---

[2] That is, pairs $s \neq t$ in S, so that $LM(f_s)$ divides $LM(f_s)$.

inclusion ambiguity could only arise if there was an $s \in S$ with $LM(f_s) = y$, and this is impossible as all elements of $\mathcal{G}$ correspond to indecomposable classes.   □

5.3 Monomial ordering and generator choice

To represent the cohomology algebra $H^*(G)$ on the computer we choose homogeneous generators $z_1, \ldots, z_n$ for $H^*(G)$ and a useable monomial ordering for the Θ-algebra $A := \Theta(z_1, \ldots, z_n)$. Then $H^*(G)$ is the quotient algebra A/I for a two-sided ideal I, and we compute a minimal Gröbner basis f_S for the *right* ideal I. As far as the computer is concerned, the cohomology ring consists of these generators, the chosen monomial ordering and this Gröbner basis.

It is well known that the size of a Gröbner basis and therefore the performance of methods using Gröbner bases depends very heavily on the chosen monomial ordering, and on the choice of generators. In this section we shall use properties of cohomology rings to develop some guidelines which at least empirically lead to suitably small Gröbner bases.

At the point when the computer has to choose generators in degree n it knows or could compute the following data:

- The cohomology groups $H^{\leq n}(G)$
- The previously chosen generators in degree $< n$
- All relations in degree $\leq n$
- The subspace of decomposable elements in $H^n(G)$
- The cohomology ring of each proper subgroup.
- The restriction map from $H^{\leq n}$ to each proper subgroup.

Conceivably we could switch to different generators later on, but this would involve a very complicated elimination of generators. So we have:

Guideline 5.1. Generators in degree n must be chosen using only information in degree $\leq n$.

5.3.1 Nilpotent generators

A theorem of Quillen says that a cohomology class is nilpotent if its restriction to every maximal elementary abelian subgroup is nilpotent. Consequently we can assume that the subspace of nilpotent classes in degree n is known when choosing new generators. The following example motivates the next guideline.

Example 5.6. Let $p > 2$ be a prime and k the field $\mathbb{F}_p$. Let F be the free graded commutative algebra $F := k[x] \otimes_k \bigwedge(y_1, y_2)$, where the generator degrees are given by $|x| = 2$ and $|y_1| = |y_2| = 1$. Let $A \subseteq F$ be the subring generated by $a := x$ and $b := y_1 y_2$. Now, A is also generated by $a = x$ and $c := x + y_1 y_2$. So we obtain two different presentations of A, namely

$$A = k[a, b]/(b^2) \quad \text{and} \quad A = k[a, c]/(c^2 - 2ac + a^2).$$

At least in this case we see that choosing nilpotent generators where possible leads to a simpler presentation.

Guideline 5.2. Each time new generators are chosen, as many nilpotent elements as possible should be chosen.

This guideline is implemented by the package Diag as follows: Let $V \subseteq H^n(G)$ be the subspace of decomposable classes and $W \subseteq H^n(G)$ the subspace of nilpotent classes. Let d be the codimension of V in H^n and $e \le d$ the codimension of V in $V + W$. Selecting new generators means choosing elements $g_1, \ldots, g_d$ of H^n which form a basis for a complement of V. In order to conform with Guideline 5.2, the computer chooses $g_1, \ldots, g_e$ in W which form a basis for a complement of V in $V + W$.

Even within these constraints the package Diag does not choose $g_1, \ldots, g_e$ at random. Recall that a cohomology class is called essential if its restriction to every proper subgroup is zero. The essential classes in $H^n(G)$ constitute a subspace of the nilpotent subspace[3], and the computer can also calculate this subspace. The program first chooses as many essential generators as possible, and only then turns to the remaining nilpotent classes. It should however be noted that essential generators are much less common than nilpotent generators: cohomology rings of 2-groups have no essential generators, and the extraspecial group of order 7^3 and exponent 7 seems to be the smallest group whose cohomology ring does have an essential indecomposable class.

Example 5.7. The algebra A is presented as follows:

$$A := k[a, b, c, d]/(c^2, cd - ab).$$

Question: What does the relation $cd = ab$ tell us? Answer: It tells us that ab is nilpotent, as c is nilpotent. So we should choose a useable ordering that makes ab the leading monomial of $cd - ab$.

Requiring $ab > cd$ does indeed lead to a simpler Gröbner basis, namely $\{c^2, ab - cd\}$. In contrast $cd > ab$ leads to the larger Gröbner basis $\{c^2, cd - ab, abc, a^2b^2\}$.

Guideline 5.3. The monomial ordering should satisfy the following condition: If b, b' are monomials of the same degree and b involves nilpotent generators but b' does not, then $b < b'$.

To comply with this guideline every monomial is assigned a "y-degree", such that every nilpotent generator has y-degree 1 and every remaining generator has y-degree zero. If b, b' are monomials of the same degree and b' has smaller y-degree than b, then we set $b < b'$.

---

[3] Here we are assuming that the group G is not elementary abelian.

5.3.2 Regular generators

Let Z be $\Omega_1(Z(G))$, the greatest central elementary abelian subgroup of G. Section 5.5 describes how to compute the restriction map from G to Z. Later on in Sect. 6.3 we shall see how to identify certain regular sequences by inspecting their restrictions to Z.

In particular, every element of $H^n(G)$ whose restriction to Z is a regular element is itself a regular element of $H^*(G)$. We shall call such classes Duflot regular elements. Regular generators can also help ensure a simple presentation of the cohomology ring. However it is undesirable to have two regular generators whose difference is a zero divisor. Ideally all regular generators should be algebraically independent of each other, but this would be difficult to guarantee in practice.

Guideline 5.4. Every opportunity to choose a completely new Duflot regular element as a generator should be acted upon.

In practice this is achieved as follows. Let $U \subseteq H^n(G)$ be the subspace of elements whose restriction to $H^n(Z)$ is nilpotent. Let $V \subseteq H^n(Z)/\sqrt{0}$ be the image under restriction of $H^n(G)$, and let $W \subseteq V$ be the image of the decomposable subspace of $H^n(G)$. First of all as many new generators as possible are chosen to lie in U. No matter how the remaining generators in this degree are chosen, their restrictions form a basis for a complement of W in V.

Guideline 5.5. Regular generators should be very small in the monomial ordering. However, Guideline 5.3 takes precedence, and so nilpotent generators should be infinitely smaller than regular generators.

A motivating example for the second part of this guideline now follows.

Example 5.8. Let a, b, y, z be elements of a cohomology ring, with y nilpotent and z Duflot regular. Assume further that the relation $by = az$ holds. Should we view this relation as a statement about by or about az? It is a statement about az, for it tells us that az is nilpotent, which means that a is nilpotent too.

To comply with the guideline, each generator x is assigned an "r-dimension". If x is Duflot regular then $\mathrm{rdim}(x) = |x|$, otherwise $\mathrm{rdim}(x) = 0$. If $b, b' \in \mathcal{B}$ have the same degree and y-degree, then we set $b < b'$ if $\mathrm{rdim}(b) > \mathrm{rdim}(b')$.

5.3.3 In summary

At last we can define the monomial ordering $\leq_{\mathrm{coho}}$ which is used in the package Diag.

Definition 5.9. *Let A be the Θ-algebra on a finite set X. Suppose that every $x \in X$ has been assigned a degree $|x|$ and in addition a y-degree $\mathrm{ydeg}(x) \in \{0, 1\}$ and a r-dimension $\mathrm{rdim}(x) \in \{0, |x|\}$. The ordering $\leq_{\mathrm{coho}}$ on $\mathcal{B}(A)$ is defined as follows. For $b, c \in \mathcal{B}(A)$ we set*

$$b \leq_{\mathrm{coho}} c :\Longleftrightarrow |b| < |c|;\ or$$
$$|b| = |c|\ and\ \mathrm{ydeg}(b) > \mathrm{ydeg}(c);\ or$$
$$|b| = |c|,\ \mathrm{ydeg}(b) = \mathrm{ydeg}(c)\ and\ \mathrm{rdim}(b) > \mathrm{rdim}(c);\ or$$
$$|b| = |c|,\ \mathrm{ydeg}(b) = \mathrm{ydeg}(c),\ \mathrm{rdim}(b) = \mathrm{rdim}(c)\ and\ b \leq_{\mathrm{lex}} c.$$

5.4 Calculating products in batches

Algorithm 5.2 calls for product coycles to be calculated in large batches. This raises the question of how most efficiently to calculate a product, and whether it is possible to acheive savings of scale. This section presents one strategy whose performance is eminently satisfactory in practice. Moreover, theoretical considerations provide a plausibility argument for this satisfactory performance. The strategy will be developed in several steps.

In order to compare the costs of various approaches it is necessary to have an estimate for the cost of computing preimages.

Assumption 5.10. Suppose we are given a map $d \colon \bigoplus^m kG \to \bigoplus^n kG$ of free kG-modules, together with a preimage Gröbner basis for d. Moreover suppose we are given r elements of the image $\mathrm{Im}(d)$. We can compute preimages of these r elements by performing Algorithm 3.37. We now have to estimate the cost of performing this algorithm.

We shall assume that the cost is directly proportional to nr, and in particular independent of m. In this section we shall say that computing these preimages costs nr units.

The product of two coycles First let $\phi \in \mathrm{H}^r(G)$ and $\psi \in \mathrm{H}^s(G)$ be cocycles. There are two ways to calculate the product $\psi.\phi$. The first method has already been described in this chapter: one constructs a lift $\phi_s \colon P_{r+s} \to P_r$, and the product is the composition $\psi \circ \phi_s$. The second method depends upon the fact that the product is graded commutative. This means that we could equally well lift ψ to $\psi_r \colon P_{r+s} \to P_r$, form the composition $\phi \circ \psi_r$ and finally multiply by $(-1)^{rs}$. Which of these two methods is better?

If the degrees r, s are equal, then there can be no difference. Now assume that $n := r + s$ is fixed. As before we shall denote by $\beta(m)$ the rank of the free kG-module P_m. Constructing ϕ_{s-i} out of ϕ_{s-i-1} involves calculating preimages of $\beta(n - i)$ elements under the map d_{s-i}. This costs $\beta(n-i).\beta(s - i - 1)$ units. It is well known that $\beta(n)$ is a function of n which grows as a polynomial whose degree is one less than the p-rank of G. For fixed n we therefore have:

1. As r increases, the number of times we have to lift ϕ in order to reach ϕ_s falls.
2. The last lift (that is, constructing ϕ_s out of ϕ_{s-1}) costs $\beta(n).\beta(n-r-1)$ units. This decreases quickly as r increases.
3. Similarly, the cost of the penultimate lift sinks fast as r increases, and so on.

We therefore formulate the following guideline:

Guideline 5.6. When we have to multiply two cocycles, we choose to lift the cocycle whose degree is larger. The bigger the difference between the two degrees becomes, the more sense this choice makes.

The product of more than two cocycles Now let $b \in \mathcal{B}(A)$ be a word of length $\ell \geq 2$ and degree n. There are then ℓ ways to factorise b as $x *_P c_x$, where x belongs to the generating set $\mathcal{G}$ and c_x to $\mathcal{B}(A)$. To be more precise there are at most ℓ such factorisations, since x^2 can only be factorised as $x *_P x$, for example. As in Algorithm 5.2 we assume that the cocycle of each c_x is known in advance. Which one of these factorisations $b = x *_P c_x$ should be used in order to calculate the cocycle of b? Guideline 5.6 prescribes the following solution:

Guideline 5.7. In this situation, one should choose an x whose degree $|x|$ is as small as possible. The product is calculated by lifting c_x.

The orderings $\leq_{\text{coho}}$ and $\leq_{\text{grlex}}$ are both based on a lexicographical ordering of $\mathcal{B}(A)$, so on an ordering on the set $\mathcal{G}$. In order to put Guideline 5.7 into practice it will be helpful to specify this ordering on $\mathcal{G}$. The ordering on $\mathcal{G}$ we shall use is the so-called chronological ordering. This means that for $z, z' \in \mathcal{G}$ we have $z < z'$ if z was added to $\mathcal{G}$ earlier than z' was. Since Algorithm 5.2 constructs $\mathcal{G}$ on a degree by degree basis, the ordering $\leq_{\text{grlex}}$ agrees on $\mathcal{G}$ with this chronological ordering. However this is not the case for the ordering $\leq_{\text{coho}}$.

To comply with Guideline 5.7 we denote by $\mathcal{G}_b$ the set

$$\mathcal{G}_b := \{x \in \mathcal{G} \mid x \text{ features in } b\}$$

and set $x(b) := \min_{\leq_{\text{grlex}}} \mathcal{G}_b$. This generator satisfies

$$|x(b)| = \min\{|x| \mid x \in \mathcal{G}_b\},$$

and so we calculate b as $x(b).c_{x(b)}$. One additional advantage of this method is that the answer always has the correct sign. This is because $x(b)$ is the first generator to occur in the word $b \in \mathcal{B}(A)$, which means that $x(b).c_{x(b)}$ does equal b rather than $-b$.

Many products in the same degree As in Algorithm 5.2, suppose we are given a set $\mathcal{P}_n$ of degree n elements of $\mathcal{B}(A)$. What is the most efficient way to compute the cocycle of each $b \in \mathcal{P}_n$? For each $b \in \mathcal{P}_n$ and for each factorisation $b = x *_P c$ with $x \in \mathcal{G}$ and $c \in \mathcal{B}(A)$ we may assume that the cocycle of c is known.

For $1 \le r \le \frac{n}{2}$ we define sets X_r, B_r and C_r as follows:

$$X_r := \{x \in \mathcal{G} \mid |x| = r\} \qquad B_r := \{b \in \mathcal{P}_n \mid x(b) \in X_r\}$$
$$C_r := \{c \in \mathcal{B}(A) \mid \text{there is a } b \in B_r \text{ with } b = x(b) *_P c\}.$$

Were $r > \frac{n}{2}$ then B_r would be the empty set. Every element of C_r has degree $n - r$.

For a single product $x *_P c$, Guideline 5.7 tells us to construct the lift c_r rather than the lift x_{n-r}. But which of the following two strategies is now more efficient?

Strategy C   Construct the rth lift c_r of each $c \in C_r$.
Strategy X   Construct the $(n - r)$th lift of each $x \in X_r$.

First we shall consider the size of the set C_r. Let $M_{n,r}$ be the size of the set $I_{n,r}$ defined by

$$I_{n,r} := \{c \in \mathcal{B}(A) \mid c \text{ is irreducible, } |c| = n - r \text{ and } |x(c)| \ge r\}.$$

As C_r is a subset of $I_{n,r}$ we have $|C_r| \le M_{n,r}$. Without the condition on $|x(c)|$ one would have $M_{n,r} = \beta(n - r)$, for all generators and relations in degree $\le n-1$ are known. Indeed we do have $M_{n,1} = \beta(n-1)$, but experience shows that $M_{n,r}/\beta(n - r)$ is a rapidly decreasing function of r.

Sticking to one of the two strategies seems to make much more sense than adopting a hybrid strategy, as we can then reuse the results of earlier calculations. For Strategy C this means that the $(r - 1)$th lift of each $c \in C_r$ and for Strategy X that the $(n - r - 1)$th lift of each $x \in X_r$ is known.

Constructing $(n - r)$th lifts of all $x \in X_r$ involves calculating $|X_r|.\beta(n)$ preimages of elements of P_{n-r-1}, at a cost of $|X_r|.\beta(n).\beta(n - r - 1)$ units. Constructing rth lifts of all $c \in C_r$ involves calculating at most $M_{n,r}.\beta(n)$ preimages of elements of P_{r-1}, at a cost of $M_{n,r}.\beta(n).\beta(r - 1)$ units. Hence

$$\text{Cost of Strategy C} = \frac{M_{n,r} \cdot \beta(r - 1)}{|X_r| \cdot \beta(n - r - 1)} \cdot \text{Cost of Strategy X}.$$

What matters is whether the rational number $\alpha_{n,r} := \frac{M_{n,r}.\beta(r-1)}{|X_r|.\beta(n-r-1)}$ is larger or smaller than one.

For $r = 1$ we have $M_{n,1} = \beta(n - 1)$ and $|X_1| = \beta(1)$. Hence $\alpha_{n,1} = \frac{\beta(n-1)}{\beta(1).\beta(n-2)}$. Assuming $\mathcal{G}$ is not cyclic then $|X_1| \ge 2$. In the extremal case $n = 2$ where both strategies are equivalent one has $\alpha_{2,1} = 1$. For large n, the ratio $\frac{\beta(n-1)}{\beta(n-2)}$ tends to one from above. So Strategy C is the winner, at least for the case $r = 1$.

The situation in the general case is

$$\alpha_{n,r} = \frac{M_{n,r} \cdot \beta(r-1)}{\beta(n-r) \cdot |X_r|} \cdot \frac{\beta(n-r)}{\beta(n-r-1)}.$$

As before $\frac{\beta(n-r)}{\beta(n-r-1)}$ tends to 1 from above, but analysing the factor $\frac{M_{n,r} \cdot \beta(r-1)}{\beta(n-r) \cdot |X_r|}$ is not so easy. As r increases, $\beta(r-1)$ increases and $\frac{M_{n,r}}{\beta(n-r)}$ decreases. For large r the size of X_r is often at most one. If X_r is empty than Strategy X has nothing to do, but Strategy C proceeds as normal.

The package Diag uses Strategy C. In practice, the case $r = 1$ always involves by far the most work – and this is the case where Strategy C is the clear winner.

Guideline 5.8. In order to calculate every $b = x(b) *_P c \in \mathcal{P}_n$ with $|x(b)| = r$, we construct the rth lift c_r of each $c \in C_r$. That is, we use Strategy C.

Remark 5.11. There is a further reason for preferring Strategy C. In practice the most time-consuming part of this strategy is the calculation of very many preimages under the map $d_1 \colon P_1 \to P_0$. However it would be very straightforward to determine the matrix of a k-linear map $s_0 \colon P_0 \to P_1$ which satisfies $d_1 s_0(x) = x$ for each $x \in \mathrm{Im}(d_1)$. Then one could compute preimages much more quickly using this matrix. This method has not yet been included in the package Diag.

It might even be worth the effort to construct the matrix of a similar splitting of d_2, in order to speed up the calculation of preimages under d_2.

Implementation of this guideline Recall that the set $I_{n,r}$ was defined by

$$I_{n,r} := \{c \in \mathcal{B}(A) \mid c \text{ is irreducible, } |c| = n - r \text{ and } |x(c)| \geq r\}.$$

The set C_r is contained in $I_{n,r}$, and for the sake of simplicity we shall lift every $c \in I_{n,r}$. The above complexity estimates allowed for this. Note moreover that $I_{n+1,r+1}$ is a subset of $I_{n,r}$, and so if c_{r-1} has been calculated for every $I_{n-1,r-1}$, then c_{r-1} is known for every $c \in I_{n,r}$.

In §5.2.1 it was shown how to construct c_r out of c_{r-1} for a single $c \in I_{n,r}$. In order to construct the lift c_r of each $c \in I_{n,r}$ in one go, one first calculates all compositions $c_{r-1} \circ d_n$ and then computes all $M_{n,r} \cdot \beta(n)$ preimages in one application of Algorithm 3.37. All lifts c_r for this pair n, r are stored in one file (the .chm file). In order to expedite the subsequent construction of c_{r+1} for $c \in I_{n+1,r+1}$, a second ordering on the set $I_{n-r} = I_{n-r+1,1}$ of all $\beta(n - r)$ irreducible monomials in degree $n - r$ is used: in this new ordering, every $c \in I_{n-r+s,s}$ appears earlier than every $c' \in I_{n-r} \setminus I_{n-r+s,s}$.

Only when the degree n part of Algorithm 5.2 has been completed is the final form of the set I_n known. To be more precise, the cocycles of the $b \in I_n$ then form a basis for $\mathrm{H}^n(G)$, and so $|I_n| = \beta(n)$. These cocycles were calculated as coordinate vectors with respect to dual basis for the k-vector space $H^n(G) = \mathrm{Hom}_{kG}\left(\bigoplus^{\beta(n)} kG, k\right) \cong \left(\bigoplus^{\beta(n)} k\right)^*$. Both basis change matrices are computed and stored (.icy file). These matrices will be useful for the computation of restriction maps. In addition the cocycles of the new generators are stored (.unz file) and their 0th lifts[4] are written down (.chm file for $(n, 0)$).

5.5 Restriction to subgroups

Let $H \neq 1$ be a subgroup of the p-group G. This section is concerned with the restriction map from the cohomology of G to that of H. The package Diag makes use of the restriction map in two different ways:

− To calculate the restriction $\mathrm{Res}_H(x)$ of each generator x of $H^*(G)$. The result should be a polynomial in the generators of $H^*(H)$. Using an elimination ordering (see Sect. 4.3) the computer can then work with the restriction map as an algebra homomorphism and determine its kernel (see Remark 4.41).
− To calculate the inverse image $\mathrm{Res}_H^{-1}(V)$ in $H^n(G)$ of a subspace $V \subseteq H^n(H)$. We have seen in Sect. 5.3 how such calculations can be used to determine the nilpotent and essential classes in $H^n(G)$. In the same way, some regular elements can be detected.

Basics Let (P_n, d_n) be the minimal resolution of the trivial kG-module k, and (Q_n, d'_n) the minimal resolution of the trivial kH-module k. As (P_n, d_n) is also a free resolution of k over kH, the inclusion map $\iota : H \to G$ lifts to a chain map

$$
\begin{array}{ccccccccccc}
\cdots \to P_n & \xrightarrow{d_n} & P_{n-1} & \to \cdots \to & P_1 & \xrightarrow{d_1} & P_0 = kG & \xrightarrow{\epsilon} & k \\
\iota_n \uparrow & & \iota_{n-1} \uparrow & & \iota_1 \uparrow & & \iota_0 \uparrow & & \| \\
\cdots \to Q_n & \xrightarrow{d'_n} & Q_{n-1} & \to \cdots \to & Q_1 & \xrightarrow{d'_1} & Q_0 = kH & \xrightarrow{\epsilon'} & k
\end{array}
\tag{5.1}
$$

The restriction $\mathrm{Res}_H(\phi)$ has as its cocycle the composition $Q_n \xrightarrow{\iota_n} P_n \xrightarrow{\phi} k$.

How subgroups are represented First of all we need to consider how to represent the inclusion map $\iota : H \hookrightarrow G$ on the computer. As described in Sect. 1.3, the group G is known to the computer as a list of permutations in the symmetric group $S_{|G|}$ which form a minimal generating set. Similarly H is represented as a subgroup of $S_{|H|}$ by a minimal generating set (h_j). For this reason the computer will view ι as a monomorphism $H \rightarrowtail G$ rather than as an inclusion $H \hookrightarrow G$.

---

[4] Recall that a cocycle is a kG-linear map $P_n \to k$, and its 0th lift is a map $P_n \to P_0 = kG$.

This monomorphism ι is represented on the computer by its matrix as a k-linear map with respect to the bases $N_{\mathbf{X}(H)}$ of kH and $N_{\mathbf{X}(G)}$ of kG. This matrix is determined by first calculating the image permutations $\iota(h_j) \in S_{|G|}$, and is stored in the .ima file.

Lifting the inclusion map Now suppose that the resolutions P_* and Q_* are known out to the nth degree, including a preimage Gröbner basis for $d_n: P_n \to P_{n-1}$. Suppose further that the $(n-1)$th lift ι_{n-1} has already been calculated, as the images in $P_{n-1} \cong \bigoplus^{\beta(n-1)} kG$ of the $\beta_H(n-1)$ free generators of $Q_{n-1} \cong \bigoplus^{\beta_H(n-1)} kH$. In particular, ι_0 is represented by $1 \in P_0 \cong kG$.

Set $M := \beta_H(n)$ and let $e_1, \ldots, e_M$ be the generators of the free kH-module Q_n. One turns $d'_n: \bigoplus^{\beta_H(n)} kH \to \bigoplus^{\beta_H(n-1)} kH$ into a map $\Delta: \bigoplus^{\beta(n)} kG \to \bigoplus^{\beta(n-1)} kG$ of free kG-modules by applying the matrix of $\iota: kN_{\mathbf{X}(H)} \to kN_{\mathbf{X}(G)}$ to each summand of $\bigoplus^{\beta_H(n-1)} kH$, which turns each $d'_n(e_i)$ into an element of $\bigoplus^{\beta(n-1)} kG$.

After this the construction proceeds in the same way as lifting a cocycle. One forms the composition $\iota_{n-1} \circ \Delta$ and sets

$$\iota_n(e_i) := \text{a preimage of } \iota_{n-1}\Delta(e_i) \text{ under } d_n.$$

Then ι_n is stored as this list of $\beta_H(n)$ elements of $\bigoplus^{\beta(n)} kG$, in the nth .icm file for H.

From this representation of ι_n the restriction map $\mathrm{Res}_H: \mathrm{H}^n(G) \to \mathrm{H}^n(H)$ can easily be recovered as its matrix with respect to the standard (i.e., dual) bases of these cohomology groups. When necessary this matrix can then be used to deduce the matrix of Res_H with respect to the bases I_n of irreducible monomials.

6 The completeness of the presentation

Chapter 5 described how to determine all generators and relations of the cohomology ring out to degree N. Carlson presented in [18] a criterion which allows one to conclude that there are no further generators or new relations in higher degrees. In the current chapter we shall recall Carlson's criterion and discuss how to put it into practice.

One part of the criterion is a condition on the Koszul complex, so the definition of this complex is recalled in Sect. 6.1. Then Carlson's criterion is stated in Sect. 6.2. Section 6.3 starts by recalling a few facts about regular sequences and then proceeds to explain how the package Diag chooses a system of parameters for the cohomology ring and calculates the depth.

If the group G has small p-rank then the cohomology of the Koszul complex and the Poincaré series of the cohomology ring are easier to determine. This is the business of Sect. 6.4, whose results were definitely known to the experts.

Carlson's criterion allows one to compute the cohomology rings of all groups of order p^n by induction in n. Section 6.5 describes how the package Diag handles cohomology rings of subgroups in order to ensure that the cohomology ring of an isomorphism class of p-groups is only calculated once.

6.1 The Koszul complex

Koszul complexes play an important role in Carlson's completeness criterion. As Carlson explains in [18], it makes sense here to view them as cochain complexes.

Hypothesis 6.1. Let $A = \bigoplus_{n \geq 0} A^n$ be a graded commutative k-algebra with $A^0 = k$, which is generated by finitely many homogeneous elements of $A^+ := \bigoplus_{n > 0} A^n$.

In particular, this hypothesis holds for the cohomology ring $\mathrm{H}^*(G)$.

Definition 6.2. *Let A be a graded commutative k-algebra as in Hypothesis 6.1.*

1. *The Koszul complex $\mathcal{K}(\zeta; A)$ of an element $\zeta \in A^n$ with $n > 0$ is by definition the following complex (C^*, δ):*
   a) *C^0 is the free A-module on one generator u_ζ.*
   b) *C^1 is the free A-module on one generator v_ζ.*
   c) *C^i is zero otherwise.*
   d) *$\delta: C^i \to C^{i+1}$ sends u_ζ to ζv_ζ (for $i = 0$).*
   Since A is itself graded the Koszul complex is bigraded: that is, we may write $C^i = \bigoplus_{j \in \mathbb{Z}} C^{ij}$ where for $x \in A^m$ we have $xu_\zeta \in C^{0,m}$ and $xv_\zeta \in C^{1,m-n}$. Then δ has bidegree $(1, 0)$.

2. *Let $\zeta_1, \ldots, \zeta_r$ be a sequence of homogeneous elements of A^+. If the prime p is odd we shall suppose further[1] that each ζ_i is an element of A^{2*}. The Koszul complex of this sequence is then defined by*

$$\mathcal{K}(\zeta_1, \ldots, \zeta_r; A) := \mathcal{K}(\zeta_1; A) \otimes_A \cdots \otimes_A \mathcal{K}(\zeta_r; A).$$

This complex inherits a bigrading and the coboundary δ has bidegree $(1, 0)$. Hence there are homology groups $\mathrm{H}^{*,j}(\mathcal{K}(\zeta_1, \ldots, \zeta_r; A))$ for each $j \in \mathbb{Z}$.

The version of Carlson's criterion in the following section differs very slightly from the original: we shall not require the parameters in Condition 6.6 to be even-dimensional if $p = 2$. The following lemma ensures that both versions of the criterion are equivalent.

Lemma 6.3. *Let A be a graded commutative k-algebra, as in Hypothesis 6.1. Let $\zeta_1, \zeta_2, \ldots, \zeta_r$ be a sequence of homogeneous elements of A^+ satisfying*

$$\mathrm{H}^{*,j}(\mathcal{K}(\zeta_1, \ldots, \zeta_r; A)) = 0 \quad \text{for every } j \geq 0. \tag{6.1}$$

Replacing ζ_1 by ζ_1^2 preserves this property (6.1).

Proof. Set $\zeta := \zeta_1$ and write $(C^{*,*}, \delta)$ for the complex $\mathcal{K}(\zeta_2, \ldots, \zeta_r; A)$. Denote by $(K^{*,*}, \delta)$ the complex $\mathcal{K}(\zeta; A)$, and write $(L^{*,*}, \delta)$ for the complex $\mathcal{K}(\zeta^2; A)$. We know that $\mathrm{H}^{*,j}(K \otimes_A C) = 0$ for every $j \geq 0$ and have to prove the same statment for $\mathrm{H}^{*,j}(L \otimes_A C)$. To simplify notation we write $u := u_\zeta$, $v := v_\zeta$, $u' := u_{\zeta^2}$, $v' := v_{\zeta^2}$ and $n := |\zeta|$.

Let $u' \otimes f + v' \otimes g$ be a cocycle in $Z^{ij}(L \otimes_A C)$ with $j \geq 0$. Hence f lies in C^{ij}, g lies in $C^{i-1,j+2n}$, and $\delta(u' \otimes f + v' \otimes g) = 0$. Consequently

$$\delta f = 0 \quad \text{and} \quad \zeta^2 f - \delta g = 0.$$

So $u \otimes \zeta f + v \otimes g \in (K \otimes_A C)^{i,j+n}$ is a cocycle and by assumption there are $a \in C^{i-1,j+n}$ and $b \in C^{i-2,j+2n}$ with $\delta(u \otimes a + v \otimes b) = u \otimes \zeta f + v \otimes g$. That is,

$$\delta a = \zeta f \quad \text{and} \quad \zeta a - \delta b = g.$$

But then $u \otimes f + v \otimes a \in (K \otimes_A C)^{ij}$ is another cocycle and bounds by assumption. So there are $c \in C^{i-1,j}$ and $e \in C^{i-2,j+m}$ satisfying $\delta(u \otimes c + v \otimes e) = u \otimes f + v \otimes a$. But then $u' \otimes c + v' \otimes (\zeta e + b)$ lies in $(L \otimes_A C)^{i-1,j}$ and satisfies $\delta(u' \otimes c + v' \otimes (\zeta e + b)) = u' \otimes f + v' \otimes g$. $\square$

---

[1] In order to avoid having to pay attention to signs.

6.2 Carlson's completeness criterion

Carlson showed in [18] that one can completely calculate the cohomology ring using only finitely many terms of the minimal resolution. His main result (his Theorem 5.6) appears here as Theorem 6.9.

Hypothesis 6.4. Let G be a finite p-group and N a natural number. As usual, k is a field of characteristic p.

   Suppose that we are given a finitely generated free graded commutative k-algebra $Q = \bigoplus_{n \geq 0} Q^n$ together with a map $\rho \colon Q \to \mathrm{H}^*(G)$ of graded commutative k-algebras and a homogeneous ideal J in Q that lies in $\mathrm{Ker}(\rho)$. In addition we shall assume the following properties:

1. Q is generated in degrees $\leq N$.
2. J is generated in degrees $\leq N$.
3. For every $n \leq N$ the map $\rho \colon Q^n \to \mathrm{H}^n(G)$ is surjective with kernel $J \cap Q^n$.

We define a graded commutative ring $R = R_N$ by $R := Q/J$ and denote by $\theta = \theta_N$ the induced homomorphism $R_N \to \mathrm{H}^*(G)$.

Remark 6.5. Given these assumptions, the map $\theta \colon R^n \to \mathrm{H}^n(G)$ is an isomorphism for each $n \leq N$. Up to isomorphism R only depends on G and N, for R is generated by the indecomposables in $\mathrm{H}^{\leq N}(G)$ together with the relations in the same degree range.

Condition 6.6 ("Condition G"). Let G, N and R be as in Hypothesis 6.4. There are $\zeta_1, \ldots, \zeta_r$ in R with the following properties:

1. $\zeta_1, \ldots, \zeta_r$ form a homogeneous system of parameters for R.
2. $\theta(\zeta_1), \ldots, \theta(\zeta_r)$ form a homogeneous system of parameters for $\mathrm{H}^*(G)$.
3. $\mathrm{H}^{*,j}(\mathcal{K}(\zeta_1, \ldots, \zeta_r; R)) = 0$ for every $j \geq 0$.
4. $\sum_{i=1}^r \max(2, |\zeta_i|) \leq N$.

If this condition holds, then r must be the p-rank of G. Property 3., which concerns the cohomology of the Koszul complex, is referred to in the Appendix as Carlson's Koszul condition

Remark 6.7. It is not yet known whether there is such a sequence in the cohomology ring of every such G. However, an existence proof for the case $r \leq 3$ has recently been given by Okuyama and Sasaki [53]. Their result holds more generally in the case $r - z \leq 2$, where z is the p-rank of the centre $Z(G)$.

Condition 6.8 ("Condition R"). For G, N, Q, J, R as in Hypothesis 6.4 set

$$\mathcal{E}_R := \bigcap \{\mathrm{Ker}(R \xrightarrow{\theta} \mathrm{H}^*(G) \xrightarrow{\mathrm{Res}} \mathrm{H}^*(H)) \mid H \leq G \text{ a maximal subgroup}\},$$

$$\mathcal{E}_Q := \bigcap \{\mathrm{Ker}(Q \xrightarrow{\rho} \mathrm{H}^*(G) \xrightarrow{\mathrm{Res}} \mathrm{H}^*(H)) \mid H \leq G \text{ a maximal subgroup}\}.$$

So $\mathcal{E}_Q$ is the inverse image in Q of $\mathcal{E}_R$. The condition requires that

1. $\mathcal{E}_Q$ is generated in degree $\leq N$.

Moreover there are homogeneous elements $y_1, \ldots, y_d$ in R which have the following properties. Here d is the p-rank of the centre $Z(G)$.

2. $y_1, \ldots, y_d$ is a regular sequence in R. (Which implies that $y_1, \ldots, y_d$ are algebraically independent over k.)
3. $\theta(y_1), \ldots, \theta(y_d)$ is a regular sequence in $\mathrm{H}^*(G)$.
4. There are finitely many homogeneous elements $\alpha_1, \ldots, \alpha_s$ of R such that $\mathcal{E}_R$ is the free module on the α_i over the polynomial algebra $k[y_1, \ldots, y_d]$. Moreover $|\alpha_i| \leq N$ for each $1 \leq i \leq s$. (Holds trivially if $\mathcal{E}_R = \{0\}$.)

Theorem 6.9 (Carlson [18]). *Let G, N, Q, J, R be as in Hypothesis 6.4. If Conditions 6.6 and 6.8 are both satisfied then $\theta \colon R \to \mathrm{H}^*(G)$ is an isomorphism.*

Proof. Let $\zeta_1, \ldots, \zeta_r$ be homogeneous elements of R satisfying Condition 6.6. Define $\bar{\zeta}_1, \ldots, \bar{\zeta}_r \in R$ by $\bar{\zeta}_i := \zeta_i^2$ if $|\zeta_i| = 1$ and $\bar{\zeta}_i := \zeta_i$ otherwise. Then $\bar{\zeta}_1, \ldots, \bar{\zeta}_r$ satisfy Condition 6.6 too, by Lemma 6.3. Moreover $|\bar{\zeta}_i| \geq 2$ for each $1 \leq i \leq r$. This reduces the theorem to the case proved by Carlson in [18]. □

6.3 Duflot regular sequences

This section begins by recalling the relevant facts about regular sequences. Then the notion of a Duflot regular sequence is introduced: a regular sequence in $\mathrm{H}^*(G)$ will be called Duflot regular if it can be proved to be regular by restricting to the centre. Such sequences are needed in order to check Condition 6.8. After this we discuss how to find a homogeneous system of parameters.

Let A be a graded commutative k-algebra, as in Hypothesis 6.1. Recall that a homogeneous element ζ of A^+ is called regular if ζ is nonzero and not a zero divisor. A sequence $\zeta_1, \ldots, \zeta_m$ of homogeneous elements of A^+ is called regular if ζ_i is a regular element of $A/(\zeta_1, \ldots \zeta_{i-1})$ for each $1 \leq i \leq m$. It easy to show that a regular sequence stays regular if the terms in the sequence are permuted. The depth of A is the length of the longest regular sequence.

Theorem 6.10. *Assuming 6.1 we have:*

1. *All maximal regular sequences in A have the same length.*
2. *The depth of A is bounded above by $\dim A$.*

If moreover A is the cohomology ring $\mathrm{H}^*(G)$ of a p-group G then:

3. *The Krull dimension $\dim \mathrm{H}^*(G)$ is equal to the p-rank r of G.*
4. *The depth τ of $\mathrm{H}^*(G)$ satisfies the inequality $z \leq \tau \leq r$, where z is the p-rank of the centre $Z(G)$.*

Proof. For parts 1. and 2. consult for example §4.3–4 of Benson's book [7] (strictly though the proof there is for the commutative case). Part 3. is a theorem of Quillen [55]. The inequality $z \leq \tau$ in Part 4. is Duflot's Theorem [28]. □

Remark 6.11. The theorem that follows allows a constructive proof of Duflot's Theorem and is also important for the computer calculation of group cohomology. It seems to have been known for some time as a Folk Theorem. The method goes back to Broto and Henn [15], who were apparently the first to consider the cohomological implications of the fact that the multiplication map $\mu: G \times C \to G$ is a group homomorphism if C is a central subgroup. Adem and Milgram [4] only proved a special case of the theorem, but their proof easily generalizes. The first explicit statement of the result in full generality is due to Carlson [17], though he describes the proof he gives as a sketch. A more detailed proof may be found in Okuyama and Sasaki's paper [53].

Theorem 6.12. ([17, 53])
Let $C \leq G$ be a central subgroup and $y_1, \ldots, y_m$ homogeneous elements of $\mathrm{H}^*(G)$. Write $\bar{y}_i$ for the restriction $\mathrm{Res}_C^G(y_i)$.

If $\bar{y}_1, \ldots, \bar{y}_m$ is a regular sequence in $\mathrm{H}^*(C)$, then $y_1, \ldots, y_m$ is a regular sequence in $\mathrm{H}^*(G)$. In particular, this happens if $\bar{y}_1, \ldots, \bar{y}_m$ is a homogeneous system of parameters for $\mathrm{H}^*(C)$ (which then implies that m is the p-rank of C). □

Definition 6.13. *A Duflot regular sequence in $\mathrm{H}^*(G)$ is a sequence of homogeneous elements of $\mathrm{H}^*(G)$ whose restrictions to $Z := \Omega_1(Z(G))$ form a regular sequence in $\mathrm{H}^*(Z)$. Here, $\Omega_1(Z(G))$ stands for the greatest central elementary abelian subgroup of G. Theorem 6.12 implies that Duflot regular sequences are indeed regular sequences. A Duflot regular sequence is called complete if its length is equal to the p-rank of Z.*

Now let $\zeta_1, \ldots, \zeta_m$ be a sequence in $\mathrm{H}^*(G)$ that might be Duflot regular. Set $n := \max\{|\zeta_i| \mid 1 \leq i \leq m\}$. In order to decide whether the sequence really is Duflot regular we only need the restrictions of the ζ_i to $Z := \Omega_1(Z(G))$. For this we only need to know $\mathrm{H}^{\leq n}(G)$, as the cohomology rings of the elementary abelian groups are well known. So we can decide whether or not a given sequence is Duflot regular without knowing the whole cohomology ring. Moreover, experience suggests that we can always choose the generators of $\mathrm{H}^*(G)$ in such a way that they contain a complete Duflot regular sequence.

Choosing a system of parameters As Carlson explains in [17, §5.3.1] it is a hard problem to choose a suitable homogeneous system of parameters, for the ground field is small and the degrees of the parameters should be as small as possible. The method currently employed by the package Diag to find a suitable system of parameters is a stop-gap. However, it is surprising how often this method does suffice.

First of all a complete Duflot regular sequence is chosen from the generators with positive r-dimension. Then the program uses rather naive methods
to try to extend this sequence to a sequence $h_1, \ldots, h_r$ with $r = p\text{-rank}(G)$
satisfying the following conditions for every maximal elementary abelian subgroup V of G:

- The restrictions $\text{Res}_V(h_1), \ldots, \text{Res}_V(h_s)$ for $s = p\text{-rank}(V)$ form a regular
  sequence in $H^*(V)$.
- $\text{Res}_V(h_i) = 0$ for $s < i \leq r$.

If such a sequence is found then it definitely is a system of parameters
for $H^*(G)$. If no such sequence can be found, or if the resulting sequence
$h_1, \ldots, h_r$ is not simultaneously a system of parameters for the current cohomology ring approximation $R = R_N$, then the cohomology computation
terminates unsuccessfully.

The depth of R is determined by finding the largest $m \leq r$ such that
$h_1, \ldots, h_m$ is a regular sequence in R. If $m < r$ then one calculates the
intersection of the annihilators $\text{Ann}_R(h_i)$ for $i > m$. If this intersection is
nontrivial then the depth of R is m.

Remark 6.14. This method for finding a system of parameters works for 50
of the 51 groups of order 32. The exceptional case is the group known to the
group library SMALL GROUPS as group number 6 of order 32. For this group
I had to find a system of parameters by hand.

Improving the selection of a system of parameters will be a priority in subsequent work on the package Diag.

6.4 Groups of small rank: Koszul complex and Poincaré series

The homology groups of the Koszul complex are easier to determine if the
p-rank of the p-group G is small. Moreover the Poincaré series can then be
easily determined from the homology of the Koszul complex.

To be more precise, the methods in this section are for the case $r - z \leq 2$,
where r is the p-rank of G and z is the p-rank of the centre $Z(G)$. These
results are definitely not new, but they are not easy to find in the literature.

Hypothesis 6.15. Let $A = \bigoplus_{n \geq 0} A^n$ be a graded commutative k-algebra with
$A^0 = k$ which is generated by finitely many homogeneous elements of $A^+ :=$
$\bigoplus_{n > 0} A^n$.

Let $\zeta_1, \zeta_2, \ldots, \zeta_r$ be a sequence of homogeneous elements of A^+, such that
(at least) the first d terms form a regular sequence in A. Write $m_j := |\zeta_j|$
and $\sigma(m) := \sum_{j=1}^r m_j$. If the prime p is odd we shall require that each m_j
be even.

Definition 6.16. *The Poincaré series $p_V(t)$ of a graded k-vector space $V = \bigoplus_{n \in \mathbb{Z}} V^n$ is defined by*

$$p_V(t) := \sum_{n \in \mathbb{Z}} t^n \dim_k(V^n).$$

Lemma 6.17. *Let A be a graded commutative k-algebra as in Hypothesis 6.15. Then*

$$p_A(t) = \frac{t^{\sigma(m)} \sum_{i=0}^{r}(-1)^{r-i} p_{H^{i,*}}(t)}{\prod_{j=1}^{r}(1 - t^{m_j})}$$

where $H^{i,j} := \mathrm{H}^{i,j}(\mathcal{K}(\zeta_1, \ldots, \zeta_r; A))$.

Proof. Write $(C^{*,*}, \delta)$ for the Koszul complex $\mathcal{K}(\zeta_1, \ldots, \zeta_r; A)$. Let $\mathcal{I}_i$ be the set of all size i subsets I of $\{1, \ldots, r\}$. Set $m_I := \sum_{j \in I} m_j$ and

$$v_I := w_{I,1} \otimes \cdots \otimes w_{I,r}, \quad \text{where} \quad w_{I,j} := \begin{cases} v_j & j \in I \\ u_j & j \notin I \end{cases}.$$

So v_I lies in $C^{i,-m_I}$.

Since $C^{i,*} = \bigoplus_{I \in \mathcal{I}_i} v_I A$ we have $p_{C^{i,*}}(t) = \sum_{I \in \mathcal{I}_i} t^{-m_I} p_A(t)$. So as the bidegree of δ is $(1, 0)$ it follows that $\sum_{i=0}^{r}(-1)^i p_{H^{i,*}}(t) = \sum_{i=0}^{r}(-1)^i p_{C^{i,*}}(t)$. $\square$

Notation 6.18. For $0 \le s \le r$ denote by $A(s)$ the quotient $A/(\zeta_1, \ldots, \zeta_s)$.

Lemma 6.19. *Suppose Hypothesis 6.15 holds with $d = r$. Set $N := \sum_{j=1}^{d} m_j$. Then*

$$p_A(t) = \frac{p_{A(r)}(t)}{\prod_{j=1}^{r}(1 - t_j^m)}.$$

Moreover $\mathrm{H}^{*,j}(\mathcal{K}(\zeta_1, \ldots, \zeta_r; A)) = 0$ if and only if $A(r)^{N+j} = 0$.

Proof. As ζ_1 is regular in A we deduce as in [18, Proposition 3.3] that

$$\mathrm{H}^{i,j}(\mathcal{K}(\zeta_1, \ldots, \zeta_r; A)) \cong \mathrm{H}^{i-1,j+m_1}(\mathcal{K}(\zeta_2, \ldots, \zeta_r; A/(\zeta_1))) \qquad (6.2)$$

The result follows by induction on r. $\square$

Lemma 6.20. *Suppose Hypothesis 6.15 holds with $d = r - 1$. Write I for the homogeneous ideal $\mathrm{Ann}_{A(r-1)}(\zeta_r)$ in $A(r-1)$. Set $N := \sum_{j=1}^{d} m_j$. Then*

$$p_A(t) = \frac{p_{A(r)}(t) - t^{m_r} p_I(t)}{\prod_{j=1}^{r}(1 - t_j^m)}.$$

Moreover $\mathrm{H}^{*,j}(\mathcal{K}(\zeta_1, \ldots, \zeta_r; A)) = 0$ if and only if the groups $A(r)^{N+m_r+j}$ and I^{N+j} are both zero.

Proof. Using Lemma 6.17 and (6.2) we reduce to the case $r = 1$, which means that $N = 0$. Then $H^{0,j} = I^j$ and $H^{1,j} = A(1)^{m_1+j}$. $\square$

Lemma 6.21. *Suppose Hypothesis 6.15 holds with $d = r - 2$. Define a homogeneous ideal $I \subseteq A(r - 2)$ by $I := \mathrm{Ann}_{A(r-2)}(\zeta_{r-1}, \zeta_r)$. Moreover define $J := \mathrm{Ann}_{A(r-2)}(\zeta_{r-1})$, $K := J/\zeta_r J$ and $L := \mathrm{Ann}_{A(r-1)}(\zeta_r)$. Set $N := \sum_{j=1}^d m_j$. Then*

$$
p_A(t) = \frac{p_{A(r)}(t) - t^{m_r} p_L(t) - t^{m_r-1} p_K(t) + t^{m_r-1+m_r} p_I(t)}{\prod_{j=1}^r (1 - t_j^m)}.
$$

Moreoever $\mathrm{H}^{*,j}(\mathcal{K}(\zeta_1, \ldots, \zeta_r; A)) = 0$ *if and only if the groups* I^{N+j}, K^{N+m_r}, $L^{N+m_{r-1}}$ *and* $A(r)^{N+m_r-1+m_r+j}$ *are all zero.*

Proof. Using Lemma 6.17 and (6.2) we reduce to the case $r = 2$, where $N = 0$. Then $H^{0,j} = I^j$ and $H^{2,j} = A(2)^{m_1+m_2+j}$. To determine $H^{1,j}$ we set

$$
S_1^j := \{(a, b) \in A^{m_1+j} \oplus A^{m_2+j} \mid \zeta_2 a = \zeta_1 b\}
$$
$$
S_2^j := \{(\zeta_1 c, \zeta_2 c) \in A^{m_1+j} \oplus A^{m_2+j} \mid c \in A^j\}
$$
$$
S_3^j := \{(0, b) \in A^{m_1+j} \oplus A^{m_2+j} \mid \zeta_1 b = 0\}.
$$

Then $H^{1,j} = S_1^j/S_2^j$ and $S_3^j \cong J^{m_2+j}$, so

$$
H^{1,j} \cong \frac{S_3^j}{S_3^j \cap S_2^j} \oplus \frac{S_1^j}{S_3^j + S_2^j}
$$
$$
\cong \frac{J^{m_2+j}}{\zeta_2 J^j} \oplus \frac{\{a \in A^{m_1+j} \mid \exists b \in A^{m_2+j} \text{ with } \zeta_2 a = \zeta_1 b\}}{\zeta_1 A^j},
$$

hence $H^{1,j} \cong K^{m_2+j} \oplus L^{m_1+j}$. $\square$

Remark 6.22. So for $r - d \leq 2$ one can calculate the homology of the Koszul complex using operations like annihilator, quotient and intersection. By contrast, for larger values of $r - d$ it seems one has to use the general method for finding the kernel of a map between free modules.

Remark 6.23. To check Condition 6.6 and calculate the Poincaré series, the computer has to solve the following task:

Let $I \subseteq J$ be right ideals in the Θ-algebra A. Determine whether the k-vector space J/I is finite-dimensional. If so, determine its Poincaré series. Gröbner bases for I and J are known.

Let $\mathcal{G}$ be the generating set of A. Let Λ_I and Λ_J be the sets of leading monomials of the Gröbner bases for I and J respectively. First one checks for each $\lambda \in \Lambda_J$ and each $x \in \mathcal{G}$ that $\lambda *_P x^N$ is divisible by at least one $\mu \in \Lambda_I$ for large N. The quotient space J/I is finite-dimensional if and only if each such pair passes this test. Then one constructs the set Q of all monomial multiples ν of elements of Λ_J such that ν is not divisible by any element of Λ_I. This set Q is a k-basis for J/I.

6.5 Identifying subgroups

To compute the cohomology ring $H^*(G)$ we have to know the restriction maps to certain subgroups. These subgroups are:

- The maximal subgroups, in order to determine the ideal of essential classes.
- The maximal elementary abelian subgroups, in order to choose nilpotent generators and a homogeneous system of parameters.
- The greatest central elementary abelian subgroup, in order to choose regular generators and identify a complete Duflot regular sequence.

We have to assume that the cohomology rings of these groups are already known, and so the computation of the cohomology rings of all p-groups of order p^n is performed by induction on n. The induction starts with the elementary abelian groups, whose cohomology is well known. More generally, the Künneth Theorem means that all generators and relations for $H^*(G)$ lie in degree at most two for any abelian G.

Isomorphism classes of small groups The Isomorphism Problem for groups of order p^n goes as follows: Decide whether or not two given groups G_1, G_2 of order p^n are isomorphic. If they are, give an isomorphism from G_1 to G_2. Calculating the cohomology ring $H^*(G)$ is a lot simpler if the computer can solve the Isomorphism Problem for subgroups of G.

Example 6.24. The extraspecial 2-group 2_+^{1+4} of order 32 has 15 maximal subgroups. Nine of these are isomorphic to $D_8 \times C_2$, and the remaining six subgroups are all isomorphic to each other. So if the computer can solve the Isomorphism Problem for groups of order 16 then it only has to perform two cohomology computations in order to known the cohomology rings of all maximal subgroups. If not then it has to perform 15 computations.

See [13, 14] for a survey of the classification problem for finite groups of small order. The Isomorphism Problem for many groups of small order is solved by the group library SMALL GROUPS, which is included in the distributions of the Computer Algebra Systems GAP [36] and MAGMA.

To be precise this group library solves the Isomorphism Problem for those p-groups whose order divides one of the numbers 2^8, 3^6, 5^5 and 7^4. So it is only for groups of order 2^{10} that one does not have a solution to the Isomorphism problem, but can hope to construct a significant portion of the minimal resolution using Diag.

The representation of inclusions In the group library, every isomorphism class is represented by a group with a preferred minimal generating set. The package Diag always uses this generating set to construct a presentation of the group algebra.

Now let H be a subgroup of G, let K be the representative of the isomorphism class of H, and let $\kappa_1, \ldots, \kappa_s$ be the preferred minimal generators of K. An isomorphism from K to H induces a monomorphism $K \rightarrowtail G$. The subgroup H is represented on the computer by specifying K together with the images in G of the κ_i. If two subgroups are isomorphic as abstract groups, then the group K will be the same in both cases but the images of the κ_i will differ.

Part III

Experimental results

7 Experimental results

In the preceding chapters a Gröbner basis method was developed for constructing minimal resolutions. Methods were developed for carrying out Carlson's approach to cohomology ring computation which

- use Gröbner bases for calculations with kG-modules, and
- work for odd primes as well as for the prime 2.

These methods were implemented in the C programming language. The package is called Diag. It uses the library of a package called "The C MeatAxe" [56] to work with vectors over $\mathbb{F}_p$, and the computer algebra system GAP [36] to assemble the necessary information about the group and its subgroup structure.

To date the applications have been in two areas:

- Computing cohomology rings of small p-groups.
- Constructing as many terms as possible in the minimal resolution for larger p-groups.

All big computations were performed on Jon F. Carlson's computer toui, a Sun ULTRA 60 Elite 3D. I am very grateful to Professor Carlson for the use of this machine.

Additionally the exact period of one periodic module was calculated to give a taste of another area where the package could be used.

7.1 Cohomology rings of small p-groups

The calculation of the mod-p cohomology rings of all groups of order p^3 was brought to completion by Leary [46] and Milgram–Tezuka [48]. The cohomology rings of all groups of order 32 were determined by Rusin [57], and Carlson recently completed the calculation of the cohomology rings of all groups of order 64 [19, 20]. For odd primes the cohomology ring of the central product $C_{p^2} * p_+^{1+2}$ of order p^4 was calculated by Benson and Carlson [10].

The package Diag was used to compute the cohomology rings of the following groups:

- All 15 groups of order $81 = 3^4$.

– 14 of the 15 groups of order $625 = 5^4$.
– Several groups of order $64 = 2^6$ and one group of order $243 = 3^5$.

To reproduce all these cohomology rings here would require well over a hundred pages. Instead, the full results have been made available on the World Wide Web, and three of the most interesting cohomology rings are described in the appendix. The address for the online cohomology rings is as follows:

    http://www.math.uni-wuppertal.de/~green/Coho/index.html

This database also includes the cohomology rings of all p-groups whose order divides one of 2^5, 3^4, 5^3 and 7^3. The groups are numbered according to the group library SMALL GROUPS (cf. Sect. 6.5).

Interpreting the tables In the rest of this section more details will be given about the individual computations. The most significant data about the cohomology rings will be presented in tabular form. The meanings of the columns will now be explained with an example.

No. Name	m	e	r	z	τ	n_E	Degrees	n_R	b_R	b_C	$|h_i|$	$\mathcal{E}$
7 $\mathrm{Syl}_3 A_9$	2	2	3	1	2	16	$1^2 2^3 3^3 4^2 5^2 6^3 7^1$	88	14	14	6,2,4	—
10	2	2	2	1	1	11	$1^2 2^3 3^1 4^1 5^1 6^2 7^1$	44	14	14	6,4	3,3,4
11 $C_9 \times 3^2$	3	2	3	3	3	6	$1^3 2^3$	3	2			

The table lists three of the 15 groups of order $81 = 3^4$. The first column says that these are groups number 7, 10 und 11 of this order. Group No. 7 is the Sylow 3-subgroup of the alternating group A_9 and group No. 11 is the direct product of the cyclic group of order 9 with the elementary abelian group of order 9. (We are using the ATLAS notation for elementary abelian groups [26].) By contrast, group No. 10 has no common name.

The subsequent columns: m is the number of minimal generators of the group, r is the p-rank of the group and z that of the centre. The exponent of the group is p^e. Column τ lists the depth of the cohomology ring. The number of minimal generators of the cohomology ring is denoted n_E, and the degrees of these generators are listed in the next column: for example, group No. 11 has three generators in degree one and three in degree two. The number of minimal relations is denoted by n_R, and b_R stands for the largest degree of a relation.

The remaining columns only concern nonabelian groups. The value of b_C is the degree in which Carlson's criterion first recognises the presentation of the cohomology ring to be complete. Then the degrees of a homogeneous system of parameters $h_1, \ldots, h_r$ is given. These parameters are ordered such that the first z terms form a complete Duflot regular sequence.

We write $\mathrm{Ess}^*(G)$ for the ideal of essential classes in $\mathrm{H}^*(G)$. That is,

$$\mathrm{Ess}^*(G) = \{x \in \mathrm{H}^*(G) \mid \mathrm{Res}_H(x) = 0 \text{ for every maximal subgroup } H < G\}.$$

So $\mathrm{Ess}^*(G) = \mathcal{E}_R$ for $R = \mathrm{H}^*(G)$ in the language of Condition 6.8 (Carlson's "Condition R"). Checking this condition involves establishing that $\mathrm{Ess}^*(G)$ is free as a module over the polynomial algebra $k[h_1, \ldots, h_z]$. The degrees of the free generators are listed in column $\mathcal{E}$. So group No. 7 has no essential classes, whereas the essential ideal for group No. 10 is free of rank three on generators in degrees $3, 3, 4$.

7.1.1 The groups of order 81

There are fifteen isomorphism classes of groups of order $81 = 3^4$. The cohomology ring of all fifteen of these groups was calculated. These computations took 12 minutes in all. Selected data on each of these cohomology rings is listed in Table 7.1.

Table 7.1. Cohomological data for the groups of order 81

No.	Name	m	e	r	z	τ	n_E	Degrees	n_R	b_R	b_C	$\lvert h_i\rvert$	$\mathcal{E}$
1	C_{81}	1	4	1	1	1	2	$1^1 2^1$	1	2			
2	$C_9 \times C_9$	2	2	2	2	2	4	$1^2 2^2$	2	2			
3		2	2	3	2	2	12	$1^2 2^4 3^2 4^1 5^1 6^2$	44	12	12	2,6,2	3,4,5
4		2	2	2	2	2	4	$1^2 2^2$	2	2	4	2,2	2
5	$C_{27} \times 3$	2	3	2	2	2	4	$1^2 2^2$	2	2			
6		2	3	2	1	1	6	$1^2 2^1 3^1 4^1 6^1$	9	10	10	6,2	2,4
7	$\mathrm{Syl}_3 A_9$	2	2	3	1	2	16	$1^2 2^3 3^3 4^2 5^2 6^3 7^1$	88	14	14	6,2,4	—
8		2	2	2	1	1	13	$1^2 2^3 3^1 4^2 5^2 6^2 7^1$	65	14	14	6,4	3
9	$\mathrm{Syl}_3 U_3(8)$	2	2	2	1	2	9	$1^2 2^4 3^2 6^1$	21	6	8	6,2	—
10		2	2	2	1	1	11	$1^2 2^3 3^1 4^1 5^1 6^2 7^1$	44	14	14	6,4	3,3,4
11	$C_9 \times 3^2$	3	2	3	3	3	6	$1^3 2^3$	3	2			
12	$3^{1+2}_+ \times 3$	3	1	3	2	3	11	$1^3 2^5 3^2 6^1$	22	6	10	2,6,2	—
13	$3^{1+2}_- \times 3$	3	2	3	2	2	8	$1^3 2^2 3^1 5^1 6^1$	10	10	10	2,6,2	3,4,5,6
14	$3^{1+2}_+ * C_9$	3	2	2	1	1	7	$1^3 2^2 4^1 6^1$	9	8	10	6,4	3
15	3^4	4	1	4	4	4	8	$1^4 2^4$	4	2			

Carlson's Depth-Essential Conjecture The following theorem is proved in [16]:

Theorem 7.1 (Carlson). *Let G be a p-group and s an integer satisfying $z < s \leq r$, where r is the p-rank of G und z that of its centre. Denote by $\mathcal{A}_s(G)$ the set of all rank s elementary abelian subgroups of G.*

If there is a class $0 \neq x \in \mathrm{H}^*(G)$ such that $\mathrm{Res}_H(x) = 0$ for the centralizer $H = C_G(V)$ of every $V \in \mathcal{A}_s(G)$, then the depth of $\mathrm{H}^*(G)$ is less than s.

In particular, this happens when G has a proper subgroup H which contains the centralizer of every $V \in \mathcal{A}_s(G)$.

Conversely the depth-essential conjecture claims that if the depth is $s-1$ then there does exist such a class x. This has been proved in the special case $s = z+1$, characterising the cases where Duflot's lower bound for the depth is tight [38]. Note that if the cohomology ring is Cohen–Macaulay[1] or contains an essential class then there is nothing to prove. Moreover it is easy to see that one only has to consider those $V \in \mathcal{A}_s(G)$ which contain the greatest central elementary abelian subgroup.

All groups of order 81 satisfy this conjecture. One sees this immediately for all apart from group No. 7, which has depth 2. But this group has precisely one rank 3 elementary abelian subgroup, and it is self-centralising. So the conditions for the last part of Theorem 7.1 are satisfied.

7.1.2 The groups of order 625

There are 15 isomorphism classes of groups of order $625 = 5^4$. To date only the cohomology rings of 14 of these groups could be computed. The computation for group No. 8 took 5 days 20 hours. Numerical data about these cohomology rings is listed in Tables 7.2 and 7.3, as the generator degrees for the cohomology rings had to be put in a separate table for space reasons.

The tables also list some information on the unfinished case, group No. 7. This is the Sylow 5-subgroup of the sporadic Conway group Co_1 and its cohomology ring is therefore of considerable interest. It is the complexity of the relations ideal that is currently preventing a complete calculation.

The data for this group are drawn from a computation out to degree 30. There are certainly at least 36 minimal generators and 556 minimal relations. It is also certain that there is a generator in degree 10 that constitutes a complete Duflot regular sequence, and that there are essential classes in degrees 4 and 5. This means that the depth of the cohomology ring is one. We see from the tables that all groups of order $625 = 5^4$ have essential classes and therefore satisfy Carlson's depth-essential conjecture.

7.1.3 Two essential classes with nonzero product

Let G be the Sylow 2-subgroup of $U_3(4)$. This is Small Group No. 245 of order 64. It group has four minimal generators, exponent 4 and p-rank 2. All involutions are central, and so the theorem of Adem and Karagueuzian [3] tells us that the cohomology ring $H^*(G)$ is Cohen–Macaulay and has essential classes. Hence the group satisfies the depth-essential conjecture. Hier are the usual cohomological data:

$r\ z\ \tau\ n_E$ Degrees $n_R\ b_R\ b_C\ |h_i|$ $\mathcal{E}$

$2\ 2\ 2\ \ 26\ \ 1^4 4^4 6^8 8^2 9^6 11^2\ \ 270\ \ 22\ \ 22\ \ 8,8\ \ 4^8 5^6 6^3 7^8 8^{16} 9^7 10^6 11^8 12^8 13^4 14^1$

---

[1] That is, the depth is equal to the Krull dimension.

Table 7.2. Cohomological data for the groups of order 625

No. Name	$m\ e\ r\ z\ \tau\ n_E$	n_R	b_R	b_C	$\lvert h_i\rvert$	$\mathcal{E}$
1 C_{625}	1 4 1 1 1 2	1	2			
2 $C_{25}\times C_{25}$	2 2 2 2 2 4	2	2			
3	2 2 3 2 2 16	90	20	20	2, 10, 2	3, 4, 5, 6, 7, 8, 9
4	2 2 2 2 2 4	2	2	4	2, 2	2
5 $C_{125}\times 5$	2 3 2 2 2 4	2	2			
6	2 3 2 1 1 8	20	18	18	10, 2	2, 4, 6, 8
7 $\mathrm{Syl}_5\,Co_1$	2 1 3 1 1 36?	556?	30?	30?	10, ?, ?	4, 5; more too?
8	2 2 3 1 1 20	151	24	24	10, 2, 4	3, 4, 5, 5, 6, 7
9	2 2 2 1 1 23	230	30	30	10, 8	3, 5, 7
10	2 2 2 1 1 23	230	30	30	10, 8	3, 5, 7
11 $C_{25}\times 5^2$	3 2 3 3 3 6	3	2			
12 $5_+^{1+2}\times 5$	3 1 3 2 2 11	51	18	18	2, 10, 6	4, 5, 5, 6, 6, 7, 7, 8
13 $5_-^{1+2}\times 5$	3 2 3 2 2 10	21	18	18	2, 10, 2	3, 4, 5, 6, 7, 8, 9, 10
14 $5_+^{1+2}*C_{25}$	3 2 2 1 1 9	22	16	16	10, 4	3, 5, 7
15 5^4	4 1 4 4 4 8	4	2			

The generator degrees are listed in Table 7.3 for space reasons.

Table 7.3. Degrees of cohomology ring generators for groups of order 625

No.	Degrees
1	$1^1 2^1$
2	$1^2 2^2$
3	$1^2 2^4 3^2 4^1 5^1 6^1 7^1 8^1 9^1 10^2$
4	$1^2 2^2$
5	$1^2 2^2$
6	$1^2 2^1 3^1 5^1 7^1 9^1 10^1$
7	$1^2 2^4 3^4 4^3 5^2 6^2 7^3 8^2 9^2 10^3 11^2 12^2 13^2 14^2 15^1$ (only up to degree 30 inclusive)
8	$1^2 2^3 3^2 4^2 5^1 6^1 7^1 8^1 9^2 10^3 11^1 12^1$
9	$1^2 2^3 3^1 4^2 5^1 6^2 7^1 8^2 9^2 10^2 11^1 12^1 13^1 14^1 15^1$
10	$1^2 2^3 3^1 4^2 5^1 6^2 7^1 8^2 9^2 10^2 11^1 12^1 13^1 14^1 15^1$
11	$1^3 2^3$
12	$1^3 2^5 3^2 7^1 8^1 9^1 10^1$
13	$1^3 2^2 3^1 5^1 7^1 9^1 10^1$
14	$1^3 2^2 4^1 6^1 8^1 10^1$
15	$1^4 2^4$

This table is a continuation of Table 7.2.

Notice how large the ideal of essential classes is. For this reason it was necessary to state the degrees of the free generators using the notation used for cohomology ring generator degrees.

This group is the first known case where the product of two essential classes is not always zero. (Of course, elementary abelian groups are excluded here.) To be more precise, the products $\text{Ess}^4(G).\text{Ess}^{10}(G)$ and $\text{Ess}^6(G).\text{Ess}^8(G)$ are one-dimensional and equal. The base of the free module $\text{Ess}^*(G)$ can be so chosen that the one 14-dimensional free generator generates these product spaces. All remaining products of essential classes are zero.

It was conjectured by Huỳnh Mui [51] and T. Marx [47] that the essential ideal squares to zero. Pham Anh Minh proved in [49] that each essential class ξ satisfies $\xi^p = 0$. Pakianathan and Yalçın related the nilpotency degree of the essential ideal to a question about fixed points of G-CW-complexes in [54]. The computation describes here disproves the Mui–Marx conjecture.

In the terminology of Hall–Senior [45] this is group No. 187 of order 64, also known as $64\Gamma_{13}a_5$. Many group invariants cannot distinguish this group from the direct product $Q_8 \times Q_8 = 64\Gamma_{10}a_3$.

This cohomology computation took 37 hours 24 minutes. More than half the time was spent on the computation and analysis of $\text{Ess}^*(G)$. It is then all the more susprising that this cohomology ring was calculated several years ago by hand, by J. Clark [25]. Using Clark's calculation one can show that the essential ideal does not square to zero without having to rely on computer calculations (see [39], which also corrects some minor errors in Clark's calculation).

Note that the cohomology rings of all remaining groups of order 64 are described on Carlson's web page [19].

7.1.4 A 3-group with Cohen–Macaulay defect 2

Let G be a p-group. As in the tables write r for the Krull dimension and τ for the depth of the cohomology ring. Most known cohomology rings are either Cohen–Macaulay ($\tau = r$), or have Cohen–Macaulay defect one. (By definition, the Cohen–Macaulay defect is the difference $r - \tau$ between Krull dimension and depth.) Previously the only exceptions were for the prime 2. We shall now meet a 3-group with defect 2.

Let G be the small group number 16 of order $243 = 3^5$. Computing the cohomology ring took 37 minutes. Here are the usual cohomological data. As there are essential classes, the depth-essential conjecture holds for this group.

m	e	r	z	τ	n_E	Degrees	n_R	b_R	b_C	$\lvert h_i \rvert$	$\mathcal{E}$
2	3	3	1	1	17	$1^2 2^3 3^2 4^2 5^2 6^3 7^2 8^1$	103	16	16	6, 2, 4	3

The Poincaré series is
$$p_G(t) = \frac{1 + 2t + 2t^2 + 2t^3 + t^4 + t^5 + 2t^6 + 2t^7 + 2t^8 + t^9}{(1 - t^2)(1 - t^4)(1 - t^6)}.$$

7.2 Resolutions for larger p-groups

The Sylow 2-subgroups of the sporadic finite simple group HS are of order 2^9, and for the Conway group Co_3 and the Mathieu group M_{24} they have order 2^{10}. A part of the minimal resolution was constructed for each of these 2-groups. The ranks as free modules of the terms in these resolutions are given in Table 7.4. There now follows a brief discussion of each case.

Table 7.4. Minimal resolutions for three Sylow 2-subgroups

Group	$Syl_2 HS$	$Syl_2 Co_3$	$Syl_2 M_{24}$
$|G|$	2^9	2^{10}	2^{10}
rkP_0	1	1	1
rkP_1	3	4	4
rkP_2	7	11	12
rkP_3	14	23	25
rkP_4	23	41	49
rkP_5	34	65	85
rkP_6	48	97	143
rkP_7	65	139	222
rkP_8	84	191	336
rkP_9	105	253	485
rkP_{10}	131	328	
rkP_{11}	163		
rkP_{12}	198		
rkP_{13}	236		
rkP_{14}	280		

Higman–Sims Let G be a Sylow 2-subgroup of HS. The 2-rank of G is 4 and its order is 2^9. The cohomology ring of this group was completely determined by Adem, Carlson, Karagueuzian and Milgram [2], based on Carlson's construction of the minimal resolution out to the 10th term. All generators and relations occur by degree 14. Using Diag the minimal resolution was constructed out to degree 14.

The third Conway group Let G be a Sylow 2-subgroup of Co_3. The 2-rank of G is 4 and its order is 2^{10}. The cohomology of Co_3 was investigated by Benson [8], but the cohomology ring of the Sylow 2-subgroup G is still not known. The package Diag was used to construct the minimal resolution out to the 10th term. In degree 8 there is a cohomology class whose restriction to the centre (cyclic of order 2) is nonzero. This class therefore constitutes a complete Duflot regular sequence. Hence it is thinkable that this cohomology ring could be completely calculated within the next few years.

The largest Mathieu group Let G be a Sylow 2-subgroup of M_{24}. The 2-rank of G is 6 and its order is 2^{10}. Maginnis has determined the cohomology ring structure out to degree 8 using spectral sequence methods. Using the package Diag I constructed the minimal resolution out to the 9th term. In degree 8 there is a cohomology class whose restriction to the centre (cyclic of order 2) is nonzero. This class therefore constitutes a complete Duflot regular sequence.

The matrix over $k = \mathbb{F}_2$ of the 8th differential d_8 would require 9,11 GB. For the construction of d_9 out of d_8 the Gröbner basis for the elimination required 130 MB, and the Gröbner basis for the kernel required 124 MB. The 9th differential itself requires 20 MB. Constructing d_9 from d_8 took 6 days and 18 hours.

7.3 The period of a periodic module

The natural 3-dimensional $\mathbb{F}_9 U_3(3)$-module M is known to be periodic [35], but its exact period is not known. The Sylow 3-subgroup G of $U_3(3)$ is extraspecial of order 27 and exponent 3. The period of M as a G-module was previously unknown too: however, by a theorem of Benson and Carlson [11], one knows that this period divides 6.

Using the package Diag I constructed the minimal resolution of M as a $\mathbb{F}_9 G$-module. As M is only defined over $\mathbb{F}_9$, the programs had to be adapted to work over non-prime fields. I was able to show that the period of the G-module M is 6. This is because the ranks as free modules of the first terms (starting with the 0th term) are $1, 2, 2, 2, 2, 1, 1, 2, 2$; and the maps d_7 and d_1 have exactly the same matrices.

For this and related computations the matrices and permutations of the Online ATLAS [60] were indispensable.

A Sample cohomology calculations

The cohomology rings of over a hundred small p-groups may be found on the World Wide Web at the address

> http://www.math.uni-wuppertal.de/~green/Coho/index.html

Some salient samples are printed in this appendix.

A.1 The cyclic group of order 2

G is C_2, the cyclic group of order 2.
This cohomology ring is well known. It has one generator: y in degree 1.
There are no relations.

A.2 The cyclic group of order 4

G is C_4, the cyclic group of order 4.
This cohomology ring is well known. It has 2 generators:

1. y in degree 1
2. x in degree 2

There is one minimal relation: $y^2 = 0$.
This minimal resolution constitutes a Gröbner basis for the ideal of relations.

A.3 The Klein 4-group

G is V_4, the elementary abelian group of order 4.
This cohomology ring is well known. It has 2 generators:

1. y_1 in degree 1
2. y_2 in degree 1

There are no relations.

A.4 The dihedral group of order 8

G is D_8, the dihedral group of order 8. It is the third Small Group of order 8 and its Hall–Senior number is 4.
G has rank 2, 2-rank 2 and exponent 4. Its centre has 2-rank 1.
The 3 maximal subgroups are: C_4, V_4 (2×).
There are 2 conjugacy classes of maximal elementary abelian subgroups. Every such subgroup has 2-rank 2.
This cohomology ring is well-known, and was successfully calculated.

Ring structure

The cohomology ring has 3 generators:

1. y_1 in degree 1
2. y_2 in degree 1
3. x in degree 2, a regular element

There is one minimal relation: $y_1 y_2 = 0$.
This minimal relation constitutes a Gröbner basis for the ideal of relations.

Essential ideal Zero ideal.

Nilradical Zero ideal.

Completion information

For this cohomology computation the minimal resolution was constructed out to degree 4. The presentation of the cohomology ring reaches its final form in degree 2. Carlson's criterion detects in degree 4 that the presentation is complete.

This cohomology ring has dimension 2 and depth 2. A homogeneous system of parameters is

$h_1 = x$ in degree 2
$h_2 = y_2^2 + y_1^2$ in degree 2

The first two terms h_1, h_2 constitute a regular sequence of maximal length. The first term h_1 constitutes a complete Duflot-regular sequence. That is to say, its restriction to the greatest central elementary abelian subgroup constitutes a regular sequence of maximal length.

Essential ideal The essential ideal is the zero ideal.

The Koszul complex A basis for $R/(h_1, h_2)$ is given below. Carlson's Koszul condition requires that each basis element has degree less than 4.

1. 1 in degree 0
2. y_2 in degree 1
3. y_1 in degree 1
4. y_1^2 in degree 2

Poincaré series $\dfrac{1+t}{(1-t)(1-t^2)}$.

Restrictions to subgroups

Restriction to maximal subgroup No. 1, isomorphic to V_4

$$y_1 \mapsto y_1 \qquad\qquad y_2 \mapsto 0 \qquad\qquad x \mapsto y_2^2 + y_1 y_2$$

Restriction to maximal subgroup No. 2, isomorphic to V_4

$$y_1 \mapsto 0 \qquad\qquad y_2 \mapsto y_1 \qquad\qquad x \mapsto y_2^2 + y_1 y_2$$

Restriction to maximal subgroup No. 3, isomorphic to C_4

$$y_1 \mapsto y \qquad\qquad y_2 \mapsto y \qquad\qquad x \mapsto x$$

Restriction to maximal elementary abelian subgroup No. 1, isomorphic to V_4

$$y_1 \mapsto 0 \qquad\qquad y_2 \mapsto y_2 + y_1 \qquad x \mapsto y_1 y_2$$

Restriction to maximal elementary abelian subgroup No. 2, isomorphic to V_4

$$y_1 \mapsto y_2 \qquad\qquad y_2 \mapsto 0 \qquad\qquad x \mapsto y_1 y_2 + y_1^2$$

Restriction to the greatest central elementary abelian subgroup, isomorphic to C_2

$$y_1 \mapsto 0 \qquad\qquad y_2 \mapsto 0 \qquad\qquad x \mapsto y^2$$

A.5 The quaternion group of order 8

G is Q_8, the quaternion group of order 8. It is the fourth Small Group of order 8 and its Hall–Senior number is 5.

G has rank 2, 2-rank 1 and exponent 4. Its centre has 2-rank 1.

The 3 maximal subgroups are: C_4 (3×).

As every involution is central, there is exactly one maximal elementary abelian subgroup. It has 2-rank 1.

This cohomology ring is well-known, and was successfully calculated.

Ring structure

The cohomology ring has 3 generators:

1. y_1 in degree 1, a nilpotent element
2. y_2 in degree 1, a nilpotent element
3. v in degree 4, a regular element

There are two minimal relations:

1. $y_2^2 = y_1 y_2 + y_1^2$
2. $y_1^3 = 0$

These minimal relations constitute a Gröbner basis for the ideal of relations.

Essential ideal There are two minimal generators:

1. $y_1 y_2$
2. y_1^2

Nilradical There are two minimal generators:

1. y_2
2. y_1

Completion information

For this cohomology computation the minimal resolution was constructed out to degree 4. The presentation of the cohomology ring reaches its final form in degree 4. Carlson's criterion detects in degree 4 that the presentation is complete.

This cohomology ring has dimension 1 and depth 1. A homogeneous system of parameters is

$h_1 = v$ in degree 4

The first term h_1 constitutes a regular sequence of maximal length. The first term h_1 constitutes a complete Duflot-regular sequence. That is to say, its restriction to the greatest central elementary abelian subgroup constitutes a regular sequence of maximal length.

Essential ideal The essential ideal is free of rank 3 as a module over the polynomial algebra in h_1. The free generators are:

$G_1 = y_1 y_2$ in degree 2
$G_2 = y_1^2$ in degree 2
$G_3 = y_1^2 y_2$ in degree 3

The essential ideal squares to zero.

The Koszul complex A basis for $R/(h_1)$ is given below. Carlson's Koszul condition requires that each basis element has degree less than 4.

1. 1 in degree 0
2. y_2 in degree 1
3. y_1 in degree 1
4. $y_1 y_2$ in degree 2
5. y_1^2 in degree 2
6. $y_1^2 y_2$ in degree 3

Poincaré series $\dfrac{1 + 2t + 2t^2 + t^3}{(1 - t^4)}$.

Restrictions to subgroups

Restriction to maximal subgroup No. 1, isomorphic to C_4

$$y_1 \mapsto 0 \qquad\qquad y_2 \mapsto y \qquad\qquad v \mapsto x^2$$

Restriction to maximal subgroup No. 2, isomorphic to C_4

$$y_1 \mapsto y \qquad\qquad y_2 \mapsto 0 \qquad\qquad v \mapsto x^2$$

Restriction to maximal subgroup No. 3, isomorphic to C_4

$$y_1 \mapsto y \qquad\qquad y_2 \mapsto y \qquad\qquad v \mapsto x^2$$

Restriction to maximal elementary abelian subgroup No. 1, isomorphic to C_2

$$y_1 \mapsto 0 \qquad\qquad y_2 \mapsto 0 \qquad\qquad v \mapsto y^4$$

Restriction to the greatest central elementary abelian subgroup, isomorphic to C_2

$$y_1 \mapsto 0 \qquad\qquad y_2 \mapsto 0 \qquad\qquad v \mapsto y^4$$

A.6 The Sylow 2-subgroup of $U_3(4)$

G is the Sylow 2-subgroup of $U_3(4)$. It is Small Group number 245 of order 64 and its Hall–Senior number is 187.

G has rank 4, 2-rank 2 and exponent 4. Its centre has 2-rank 2.

All fifteen maximal subgroups are isomorphic to the Small Group number 32 of order 32.

As every involution is central, there is exactly one maximal elementary abelian subgroup. It has 2-rank 2.

This cohomology ring was calculated by Clark [25], albeit with minor errors (see [39]). The computer calculation was successful.

Ring structure

The cohomology ring has 26 generators:

1. y_1 in degree 1, a nilpotent element
2. y_2 in degree 1, a nilpotent element
3. y_3 in degree 1, a nilpotent element
4. y_4 in degree 1, a nilpotent element
5. v_1 in degree 4, a nilpotent element
6. v_2 in degree 4, a nilpotent element
7. v_3 in degree 4, a nilpotent element
8. v_4 in degree 4, a nilpotent element
9. t_1 in degree 6, a nilpotent element
10. t_2 in degree 6, a nilpotent element
11. t_3 in degree 6, a nilpotent element
12. t_4 in degree 6, a nilpotent element
13. t_5 in degree 6, a nilpotent element
14. t_6 in degree 6, a nilpotent element
15. t_7 in degree 6, a nilpotent element
16. t_8 in degree 6, a nilpotent element
17. r_1 in degree 8, a regular element
18. r_2 in degree 8, a regular element
19. q_1 in degree 9, a nilpotent element
20. q_2 in degree 9, a nilpotent element
21. q_3 in degree 9, a nilpotent element
22. q_4 in degree 9, a nilpotent element
23. q_5 in degree 9, a nilpotent element
24. q_6 in degree 9, a nilpotent element
25. o_1 in degree 11, a nilpotent element
26. o_2 in degree 11, a nilpotent element

There are 270 minimal relations:

1. $y_4^2 = y_2y_4 + y_2y_3 + y_1y_4 + y_1y_2 + y_1^2$
2. $y_3^2 = y_2y_3 + y_2^2 + y_1y_4 + y_1y_3$
3. $y_2^2y_3 = y_1y_2y_4 + y_1y_2y_3 + y_1y_2^2 + y_1^2y_4$
4. $y_2^3 = y_1y_2y_3 + y_1y_2^2 + y_1^2y_4 + y_1^2y_2 + y_1^3$
5. $y_4v_4 = y_2v_4 + y_2v_2 + y_1v_3 + y_1v_2 + y_1v_1 + y_1^3y_3y_4 + y_1^3y_2y_4 + y_1^4y_4 + y_1^4y_2 + y_1^5$
6. $y_4v_3 = y_2v_2 + y_1v_4 + y_1v_3 + y_1v_2 + y_1^4y_3 + y_1^5$
7. $y_4v_2 = y_2v_4 + y_1v_1 + y_1^3y_3y_4 + y_1^3y_2y_4 + y_1^4y_2 + y_1^5$
8. $y_4v_1 = y_2v_3 + y_1v_1 + y_1^3y_3y_4 + y_1^4y_3 + y_1^4y_2 + y_1^5$
9. $y_3v_4 = y_2v_4 + y_2v_3 + y_1v_3 + y_1^3y_3y_4 + y_1^4y_2$
10. $y_3v_3 = y_2v_4 + y_2v_2 + y_1v_4 + y_1^4y_4 + y_1^4y_3 + y_1^4y_2$
11. $y_3v_2 = y_2v_2 + y_1v_4 + y_1v_1 + y_1^4y_3 + y_1^4y_2 + y_1^5$
12. $y_3v_1 = y_2v_2 + y_1v_3 + y_1v_2 + y_1v_1 + y_1^3y_2y_4 + y_1^4y_3 + y_1^5$
13. $y_2v_1 = y_1v_2 + y_1^4y_3 + y_1^4y_2$

14. $y_1^3 y_2 y_3 = y_1^4 y_4 + y_1^4 y_2$

15. $y_1^3 y_2^2 = y_1^4 y_2$

16. $y_4 t_8 = y_2 t_8 + y_2 t_4 + y_1 t_6 + y_1 t_3 + y_1 t_1 + y_1 y_2^2 v_4 + y_1^2 y_2 v_2 + y_1^3 v_4 + y_1^3 v_3 + y_1^3 v_2 + y_1^3 v_1$

17. $y_4 t_7 = y_2 t_8 + y_2 t_3 + y_1 t_6 + y_1 y_2^2 v_4 + y_1^3 v_3 + y_1^3 v_2 + y_1^3 v_1$

18. $y_4 t_6 = y_2 t_4 + y_1 t_7 + y_1 t_5 + y_1 t_3 + y_1 t_1 + y_1^2 y_2 v_4 + y_1^3 v_4 + y_1^3 v_2$

19. $y_4 t_5 = y_2 t_8 + y_1 t_8 + y_1 t_4 + y_1 t_1 + y_1^2 y_2 v_4 + y_1^2 y_2 v_3 + y_1^3 v_4 + y_1^3 v_2$

20. $y_4 t_4 = y_2 t_8 + y_2 t_7 + y_2 t_4 + y_2 t_3 + y_1 t_8 + y_1 t_6 + y_1 t_5 + y_1 t_4 + y_1 t_3 + y_1 t_2 + y_1^2 y_2 v_3 + y_1^3 v_4 + y_1^3 v_2$

21. $y_4 t_3 = y_2 t_7 + y_2 t_3 + y_1 t_8 + y_1 t_7 + y_1 t_6 + y_1 t_5 + y_1 t_2 + y_1 y_2^2 v_4 + y_1^2 y_2 v_4 + y_1^2 y_2 v_3 + y_1^3 v_4 + y_1^3 v_2 + y_1^3 v_1$

22. $y_4 t_2 = y_2 t_4 + y_1 t_6 + y_1 t_4 + y_1 t_3 + y_1^3 v_3$

23. $y_4 t_1 = y_2 t_3 + y_1 t_5 + y_1 t_4 + y_1 t_2 + y_1 y_2^2 v_4 + y_1^2 y_2 v_4 + y_1^3 v_4$

24. $y_3 t_8 = y_2 t_8 + y_2 t_7 + y_2 t_4 + y_2 t_3 + y_1 t_8 + y_1 t_7 + y_1 t_6 + y_1 t_5 + y_1 t_4 + y_1 t_3 + y_1 t_1 + y_1 y_2^2 v_4 + y_1^3 v_4 + y_1^3 v_1$

25. $y_3 t_7 = y_2 t_8 + y_2 t_3 + y_1 t_7 + y_1 t_5 + y_1 t_4 + y_1 t_3 + y_1 t_2 + y_1^3 v_4$

26. $y_3 t_6 = y_2 t_3 + y_1 t_8 + y_1 t_6 + y_1 t_5 + y_1 t_3 + y_1 t_2 + y_1 y_2^2 v_4 + y_1^2 y_2 v_2 + y_1^3 v_4 + y_1^3 v_2 + y_1^3 v_1$

27. $y_3 t_5 = y_2 t_4 + y_2 t_3 + y_1 t_7 + y_1 t_3 + y_1 t_2 + y_1 t_1 + y_1 y_2^2 v_4 + y_1^2 y_2 v_3 + y_1^2 y_2 v_2 + y_1^3 v_4 + y_1^3 v_2 + y_1^3 v_1$

28. $y_3 t_4 = y_2 t_4 + y_2 t_3 + y_1 t_6 + y_1 t_3 + y_1^2 y_2 v_3 + y_1^3 v_4$

29. $y_3 t_3 = y_2 t_4 + y_1 t_5 + y_1 t_2 + y_1 y_2^2 v_4 + y_1^3 v_4 + y_1^3 v_2 + y_1^3 v_1$

30. $y_3 t_2 = y_1 t_5 + y_1 y_2^2 v_4 + y_1^2 y_2 v_3 + y_1^2 y_2 v_2 + y_1^3 v_4$

31. $y_3 t_1 = y_1 t_5 + y_1 t_4 + y_1 t_3 + y_1 y_2^2 v_4 + y_1^2 y_2 v_4 + y_1^2 y_2 v_3 + y_1^3 v_1$

32. $y_2 t_6 = y_2 t_4 + y_2 t_3 + y_1 t_6 + y_1 t_4 + y_1 t_3 + y_1 t_1 + y_1 y_2^2 v_4 + y_1^3 v_3$

33. $y_2 t_5 = y_2 t_4 + y_1 t_6 + y_1 t_5 + y_1 t_3 + y_1^2 y_2 v_4 + y_1^2 y_2 v_3 + y_1^2 y_2 v_2 + y_1^3 v_4 + y_1^3 v_3 + y_1^3 v_1$

34. $y_2 t_2 = y_1 t_5 + y_1 t_4 + y_1 t_2 + y_1 t_1 + y_1^3 v_4$

35. $y_2 t_1 = y_1 t_2 + y_1 t_1 + y_1^2 y_2 v_4 + y_1^2 y_2 v_3 + y_1^3 v_4 + y_1^3 v_3 + y_1^3 v_1$

36. $v_4^2 = 0$

37. $v_3 v_4 = y_1^3 y_2 v_4 + y_1^3 y_2 v_3 + y_1^4 v_4 + y_1^4 v_3$

38. $v_3^2 = 0$

39. $v_2 v_4 = y_1^3 y_2 v_2 + y_1^4 v_4 + y_1^4 v_3 + y_1^4 v_2 + y_1^4 v_1$

40. $v_2 v_3 = y_1^3 y_2 v_3 + y_1^4 v_1$

41. $v_2^2 = 0$

42. $v_1 v_4 = y_1^2 y_2^2 v_4 + y_1^3 y_2 v_2 + y_1^4 v_4 + y_1^4 v_2$

43. $v_1 v_3 = y_1^3 y_2 v_4 + y_1^3 y_2 v_2 + y_1^4 v_4$

44. $v_1 v_2 = y_1^3 y_2 v_3 + y_1^3 y_2 v_2 + y_1^4 v_4 + y_1^4 v_3$

45. $v_1^2 = 0$

46. $y_1^3 t_5 = y_1^3 t_4 + y_1^3 t_2 + y_1^3 t_1$

47. $v_4 t_8 = y_2 q_4 + y_2 q_3 + y_1 q_6 + y_1 q_5 + y_2^2 r_2 + y_2^2 r_1 + y_1 y_3 r_2 + y_1 y_3 r_1 + y_1 y_2 r_2 + y_1^2 r_2 + y_1^2 r_1 + y_1^4 t_2 + y_1^4 t_1$

48. $v_4 t_7 = y_2 q_4 + y_2 q_3 + y_1 q_6 + y_1 q_2 + y_1 q_1 + y_2^2 r_2 + y_2^2 r_1 + y_1 y_4 r_2 + y_1 y_3 r_2 + y_1 y_3 r_1 + y_1 y_2 r_2 + y_1 y_2 r_1 + y_1^2 r_2 + y_1^2 r_1 + y_1^4 t_4$

49. $v_4t_6 = y_2q_3 + y_1q_3 + y_1q_2 + y_2y_4r_2 + y_2y_4r_1 + y_2^2r_1 + y_1y_4r_2 + y_1y_4r_1 + y_1y_2r_1 + y_1^4t_4 + y_1^4t_3 + y_1^4t_2 + y_1^4t_1$

50. $v_4t_5 = y_2q_4 + y_1q_5 + y_1q_3 + y_1q_2 + y_1q_1 + y_2y_4r_2 + y_2y_4r_1 + y_2^2r_2 + y_1y_4r_1 + y_1y_2r_2 + y_1y_2r_1 + y_1^4t_4 + y_1^4t_3 + y_1^4t_2 + y_1^4t_1$

51. $v_4t_4 = y_1q_6 + y_1q_5 + y_1q_4 + y_1q_3 + y_1y_3r_2 + y_1y_3r_1 + y_1y_2r_2 + y_1^4t_2$

52. $v_4t_3 = y_2q_4 + y_2q_3 + y_1q_5 + y_1q_4 + y_1q_3 + y_2^2r_2 + y_2^2r_1 + y_1y_4r_2 + y_1y_4r_1 + y_1y_3r_2 + y_1y_2r_1 + y_1^2r_2 + y_1^2r_1 + y_1^4t_3 + y_1^4t_2$

53. $v_4t_2 = y_2q_3 + y_1q_6 + y_1q_5 + y_1q_2 + y_1q_1 + y_2y_4r_2 + y_2y_4r_1 + y_2^2r_1 + y_1y_4r_1 + y_1y_3r_1 + y_1y_2r_1 + y_1^2r_2 + y_1^2r_1 + y_1^4t_4 + y_1^4t_2 + y_1^4t_1$

54. $v_4t_1 = y_2q_3 + y_1q_6 + y_1q_5 + y_1q_4 + y_2y_4r_2 + y_2y_4r_1 + y_2^2r_1 + y_1y_4r_2 + y_1y_4r_1 + y_1y_3r_2 + y_1y_3r_1 + y_1y_2r_2 + y_1y_2r_1 + y_1^4t_1$

55. $v_3t_8 = y_2q_3 + y_1q_5 + y_2y_4r_2 + y_2y_4r_1 + y_2^2r_1 + y_1y_4r_2 + y_1y_4r_1 + y_1y_3r_2 + y_1y_2r_1 + y_1^4t_3$

56. $v_3t_7 = y_2q_4 + y_2q_3 + y_1q_4 + y_1q_3 + y_2^2r_2 + y_2^2r_1 + y_1^2r_2 + y_1^2r_1 + y_1^4t_4 + y_1^4t_3 + y_1^4t_2 + y_1^4t_1$

57. $v_3t_6 = y_1q_6 + y_1q_5 + y_1q_4 + y_1q_2 + y_1q_1 + y_1y_4r_2 + y_1y_3r_1 + y_1y_2r_2 + y_1y_2r_1 + y_1^4t_4 + y_1^4t_3 + y_1^4t_2 + y_1^4t_1$

58. $v_3t_5 = y_2q_4 + y_2q_3 + y_1q_6 + y_1q_4 + y_1q_1 + y_2^2r_2 + y_2^2r_1 + y_1y_4r_1 + y_1y_3r_2 + y_1y_3r_1 + y_1y_2r_1 + y_1^4t_1$

59. $v_3t_4 = y_2q_4 + y_2q_3 + y_1q_6 + y_1q_5 + y_2^2r_2 + y_2^2r_1 + y_1y_3r_2 + y_1y_3r_1 + y_1y_2r_2 + y_1^2r_2 + y_1^2r_1 + y_1^4t_4 + y_1^4t_3 + y_1^4t_2 + y_1^4t_1$

60. $v_3t_3 = y_2q_3 + y_1q_6 + y_1q_4 + y_2y_4r_2 + y_2y_4r_1 + y_2^2r_1 + y_1y_3r_1 + y_1y_2r_2 + y_1^4t_4$

61. $v_3t_2 = y_2q_3 + y_1q_6 + y_1q_5 + y_1q_4 + y_1q_3 + y_2y_4r_2 + y_2y_4r_1 + y_2^2r_1 + y_1y_3r_2 + y_1y_3r_1 + y_1y_2r_2 + y_1^4t_4$

62. $v_3t_1 = y_1q_6 + y_1q_4 + y_1q_3 + y_1q_2 + y_1y_4r_2 + y_1y_4r_1 + y_1y_3r_1 + y_1y_2r_2 + y_1y_2r_1 + y_1^4t_4 + y_1^4t_3 + y_1^4t_2 + y_1^4t_1$

63. $v_2t_8 = y_2q_4 + y_1q_6 + y_1q_5 + y_1q_3 + y_2y_4r_2 + y_2y_4r_1 + y_2^2r_2 + y_1y_4r_2 + y_1y_4r_1 + y_1y_3r_2 + y_1y_3r_1 + y_1y_2r_2 + y_1y_2r_1 + y_1^2r_2 + y_1^2r_1 + y_1^4t_4 + y_1^4t_1$

64. $v_2t_7 = y_2q_4 + y_1q_6 + y_1q_5 + y_1q_2 + y_1q_1 + y_2y_4r_2 + y_2y_4r_1 + y_2^2r_2 + y_1y_4r_1 + y_1y_3r_1 + y_1y_2r_2 + y_1^2r_2 + y_1^2r_1 + y_1^4t_4 + y_1^4t_1$

65. $v_2t_6 = y_1q_5 + y_1q_4 + y_1q_2 + y_1q_1 + y_1y_4r_1 + y_1y_2r_2 + y_1y_2r_1 + y_1^2r_2 + y_1^2r_1 + y_1^4t_2$

66. $v_2t_5 = y_2q_3 + y_1q_5 + y_1q_4 + y_1q_2 + y_1q_1 + y_2y_4r_2 + y_2y_4r_1 + y_2^2r_1 + y_1y_4r_1 + y_1y_2r_2 + y_1y_2r_1 + y_1^2r_2 + y_1^2r_1 + y_1^4t_4 + y_1^4t_2 + y_1^4t_1$

67. $v_2t_4 = y_2q_3 + y_1q_6 + y_1q_2 + y_1q_1 + y_2y_4r_2 + y_2y_4r_1 + y_2^2r_1 + y_1y_4r_2 + y_1y_3r_2 + y_1y_3r_1 + y_1^2r_2 + y_1^2r_1 + y_1^4t_3$

68. $v_2t_3 = y_2q_3 + y_1q_2 + y_1q_1 + y_2y_4r_2 + y_2y_4r_1 + y_2^2r_1 + y_1y_4r_1 + y_1y_3r_2 + y_1^4t_4 + y_1^4t_3 + y_1^4t_2 + y_1^4t_1$

69. $v_2t_2 = y_1q_5 + y_1q_1 + y_1y_4r_2 + y_1y_2r_1 + y_1^4t_3 + y_1^4t_2$

70. $v_2t_1 = y_1q_5 + y_1q_2 + y_1q_1 + y_1y_4r_2 + y_1y_2r_1 + y_1^4t_3 + y_1^4t_2 + y_1^4t_1$

71. $v_1t_8 = y_2q_4 + y_1q_4 + y_1q_2 + y_2y_4r_2 + y_2y_4r_1 + y_2^2r_2 + y_1y_4r_2 + y_1y_4r_1 + y_1y_2r_1 + y_1^2r_2 + y_1^2r_1 + y_1^4t_4 + y_1^4t_2 + y_1^4t_1$

72. $v_1t_7 = y_2q_3 + y_1q_6 + y_1q_5 + y_1q_2 + y_2y_4r_2 + y_2y_4r_1 + y_2^2r_1 + y_1y_3r_2 + y_1y_3r_1 + y_1y_2r_1 + y_1^2r_2 + y_1^2r_1 + y_1^4t_4$

73. $v_1t_6 = y_2q_3 + y_1q_1 + y_2y_4r_2 + y_2y_4r_1 + y_2^2r_1 + y_1y_4r_1 + y_1y_3r_2 + y_1^4t_3 + y_1^4t_1$

74. $v_1t_5 = y_2q_3 + y_1q_3 + y_1q_1 + y_2y_4r_2 + y_2y_4r_1 + y_2^2r_1 + y_1y_4r_2 + y_1y_3r_2 + y_1y_2r_1 + y_1^4t_4 + y_1^4t_3$

75. $v_1t_4 = y_2q_3 + y_1q_3 + y_1q_2 + y_1q_1 + y_2y_4r_2 + y_2y_4r_1 + y_2^2r_1 + y_1y_4r_2 + y_1y_3r_2 + y_1y_2r_1 + y_1^4t_2 + y_1^4t_1$

76. $v_1t_3 = y_1q_6 + y_1q_5 + y_1q_4 + y_1q_3 + y_1y_3r_2 + y_1y_3r_1 + y_1y_2r_2 + y_1^4t_4 + y_1^4t_2$

77. $v_1t_2 = y_1q_5 + y_1q_2 + y_1y_4r_2 + y_1y_4r_1 + y_1y_3r_2 + y_1y_2r_1 + y_1^4t_3$

78. $v_1t_1 = y_1q_1 + y_1y_4r_1 + y_1y_3r_2 + y_1^4t_3 + y_1^4t_2 + y_1^4t_1$

79. $y_4q_6 = y_2q_4 + y_2q_3 + y_1q_5 + y_1q_3 + y_1q_1 + y_3y_4r_1 + y_2y_4r_2 + y_2y_4r_1 + y_2y_3r_2 + y_2y_3r_1 + y_2^2r_2 + y_2^2r_1 + y_1y_4r_1 + y_1^2r_2 + y_1^2r_1 + y_1^4t_3 + y_1^4t_1$

80. $y_4q_5 = y_2q_4 + y_2q_3 + y_1q_6 + y_1q_1 + y_3y_4r_2 + y_2y_4r_2 + y_2y_3r_2 + y_2y_3r_1 + y_2^2r_2 + y_2^2r_1 + y_1y_4r_1 + y_1y_3r_2 + y_1y_3r_1 + y_1^4t_3 + y_1^4t_2 + y_1^4t_1$

81. $y_4q_4 = y_1q_6 + y_1q_5 + y_1q_4 + y_1q_2 + y_1q_1 + y_2y_4r_1 + y_2y_3r_2 + y_2y_3r_1 + y_1y_4r_2 + y_1y_3r_1 + y_1^2r_2 + y_1^2r_1 + y_1^4t_3 + y_1^4t_2$

82. $y_4q_3 = y_1q_5 + y_1q_1 + y_2y_4r_2 + y_2y_3r_2 + y_2y_3r_1 + y_1y_4r_1 + y_1y_2r_2 + y_1^2r_2 + y_1^2r_1 + y_1^4t_2$

83. $y_4q_2 = y_2q_4 + y_1q_6 + y_1q_5 + y_1q_4 + y_2y_4r_2 + y_2y_4r_1 + y_2^2r_2 + y_1y_4r_2 + y_1y_4r_1 + y_1y_3r_2 + y_1y_3r_1 + y_1^4t_1$

84. $y_4q_1 = y_2q_3 + y_1q_5 + y_1q_2 + y_1q_1 + y_3y_4r_2 + y_2y_4r_2 + y_2y_3r_1 + y_2^2r_1 + y_1y_4r_2 + y_1y_4r_1 + y_1^2r_1 + y_1^4t_4 + y_1^4t_2$

85. $y_3q_6 = y_1q_6 + y_1q_5 + y_1q_1 + y_3y_4r_2 + y_3y_4r_1 + y_2y_3r_1 + y_2^2r_1 + y_1y_3r_2 + y_1y_3r_1 + y_1y_2r_1 + y_1^2r_2 + y_1^2r_1 + y_1^4t_3$

86. $y_3q_5 = y_2q_3 + y_1q_6 + y_1q_3 + y_1q_1 + y_3y_4r_2 + y_3y_4r_1 + y_2y_4r_2 + y_2y_4r_1 + y_2y_3r_2 + y_2y_3r_1 + y_2^2r_2 + y_2^2r_1 + y_1y_4r_2 + y_1y_4r_1 + y_1y_3r_1 + y_1y_2r_1 + y_1^2r_2 + y_1^2r_1 + y_1^4t_4 + y_1^4t_2$

87. $y_3q_4 = y_2q_4 + y_2q_3 + y_1q_6 + y_1q_5 + y_1q_4 + y_1q_3 + y_1q_2 + y_3y_4r_2 + y_3y_4r_1 + y_2y_3r_2 + y_2^2r_2 + y_2^2r_1 + y_1y_2r_1 + y_1^4t_3$

88. $y_3q_3 = y_2q_4 + y_1q_5 + y_1q_4 + y_1q_3 + y_1q_2 + y_3y_4r_2 + y_3y_4r_1 + y_2y_4r_2 + y_2y_4r_1 + y_2y_3r_1 + y_2^2r_2 + y_1y_4r_2 + y_1y_4r_1 + y_1y_3r_2 + y_1y_2r_1 + y_1^2r_2 + y_1^2r_1 + y_1^4t_4 + y_1^4t_3 + y_1^4t_2 + y_1^4t_1$

89. $y_3q_2 = y_1q_6 + y_1y_4r_2 + y_1y_4r_1 + y_1y_3r_1 + y_1^2r_2 + y_1^2r_1 + y_1^4t_4 + y_1^4t_3 + y_1^4t_1$

90. $y_3q_1 = y_1q_6 + y_1q_5 + y_1q_4 + y_1q_3 + y_1q_2 + y_3y_4r_1 + y_2y_3r_2 + y_2^2r_2 + y_1y_4r_2 + y_1y_3r_1 + y_1y_2r_2 + y_1^4t_4 + y_1^4t_3 + y_1^4t_2 + y_1^4t_1$

91. $y_2q_6 = y_2q_4 + y_1q_6 + y_1q_4 + y_1q_2 + y_2y_3r_1 + y_2^2r_2 + y_1y_3r_1 + y_1y_2r_2 + y_1^4t_3 + y_1^4t_2 + y_1^4t_1$

92. $y_2q_5 = y_1q_6 + y_1q_4 + y_1q_1 + y_2y_4r_2 + y_2y_4r_1 + y_2y_3r_2 + y_2^2r_1 + y_1y_4r_1 + y_1y_3r_2 + y_1y_3r_1 + y_1y_2r_2 + y_1^4t_4 + y_1^4t_3 + y_1^4t_2 + y_1^4t_1$

93. $y_2q_2 = y_1q_6 + y_1q_5 + y_1q_4 + y_1y_4r_2 + y_1y_4r_1 + y_1y_3r_2 + y_1y_3r_1 + y_1y_2r_2 + y_1y_2r_1 + y_1^4t_4 + y_1^4t_3 + y_1^4t_1$

94. $y_2q_1 = y_1q_5 + y_1q_2 + y_1q_1 + y_2y_4r_1 + y_2y_3r_2 + y_1y_4r_2 + y_1y_2r_1 + y_1^4t_4 + y_1^4t_2 + y_1^4t_1$

95. $t_8^2 = y_1^2y_2y_4r_2 + y_1^2y_2y_3r_2 + y_1^2y_2y_3r_1 + y_1^2y_2^2r_1 + y_1^3y_4r_2 + y_1^3y_4r_1 + y_1^3y_3r_1 + y_1^3y_2r_2$

96. $t_7t_8 = y_1^2y_2q_4 + y_1^3q_6 + y_1^3q_5 + y_1^3q_4 + y_1^3q_1 + y_1^2y_3y_4r_2 + y_1^2y_3y_4r_1 + y_1^2y_2y_4r_1 + y_1^2y_2y_3r_1 + y_1^3y_4r_1 + y_1^3y_3r_1 + y_1^3y_2r_1 + y_1^4r_1$

97. $t_7^2 = y_1^2 y_2 y_4 r_1 + y_1^2 y_2 y_3 r_2 + y_1^2 y_2^2 r_2 + y_1^2 y_2^2 r_1 + y_1^3 y_4 r_2 + y_1^3 y_3 r_2 + y_1^3 y_3 r_1 + y_1^3 y_2 r_1$

98. $t_6 t_8 = y_1^3 q_6 + y_1^3 q_5 + y_1^3 q_3 + y_1^2 y_3 y_4 r_2 + y_1^2 y_2 y_4 r_2 + y_1^2 y_2 y_4 r_1 + y_1^2 y_2 y_3 r_2 + y_1^2 y_2^2 r_2 + y_1^3 y_3 r_2 + y_1^3 y_2 r_2 + y_1^4 r_2$

99. $t_6 t_7 = y_1^2 y_2 q_4 + y_1^3 q_6 + y_1^3 q_4 + y_1^3 q_2 + y_1^2 y_3 y_4 r_1 + y_1^2 y_2 y_4 r_1 + y_1^3 y_4 r_2 + y_1^3 y_4 r_1 + y_1^3 y_3 r_2 + y_1^3 y_2 r_2 + y_1^4 r_2 + y_1^4 r_1$

100. $t_6^2 = y_1^2 y_2 y_4 r_2 + y_1^2 y_2 y_3 r_2 + y_1^2 y_2 y_3 r_1 + y_1^2 y_2^2 r_2 + y_1^2 y_2^2 r_1 + y_1^3 y_4 r_2 + y_1^3 y_4 r_1 + y_1^3 y_3 r_1 + y_1^3 y_2 r_2 + y_1^4 r_1$

101. $t_5 t_8 = y_1^2 y_2 q_4 + y_1^3 q_6 + y_1^3 q_3 + y_1^3 q_2 + y_1^3 q_1 + y_1^2 y_2 y_4 r_2 + y_1^2 y_2 y_3 r_2 + y_1^2 y_2^2 r_1 + y_1^3 y_4 r_2 + y_1^3 y_4 r_1 + y_1^3 y_2 r_2 + y_1^3 y_2 r_1 + y_1^4 r_2 + y_1^4 r_1$

102. $t_5 t_7 = y_1^2 y_2 q_4 + y_1^3 q_6 + y_1^3 q_5 + y_1^3 q_4 + y_1^3 q_3 + y_1^2 y_3 y_4 r_2 + y_1^2 y_2 y_4 r_2 + y_1^2 y_2 y_3 r_2 + y_1^2 y_2 y_3 r_1 + y_1^2 y_2^2 r_2 + y_1^2 y_2^2 r_1 + y_1^3 y_4 r_2 + y_1^3 y_3 r_2 + y_1^3 y_3 r_1 + y_1^3 y_2 r_2 + y_1^4 r_2 + y_1^4 r_1$

103. $t_5 t_6 = y_1^2 y_2 q_4 + y_1^3 q_5 + y_1^3 q_4 + y_1^3 q_3 + y_1^3 q_2 + y_1^2 y_3 y_4 r_1 + y_1^2 y_2 y_4 r_1 + y_1^2 y_2 y_3 r_1 + y_1^2 y_2^2 r_2 + y_1^2 y_2^2 r_1 + y_1^3 y_4 r_1 + y_1^3 y_3 r_1$

104. $t_5^2 = y_1^2 y_2 y_4 r_2 + y_1^2 y_2 y_4 r_1 + y_1^2 y_2 y_3 r_1 + y_1^2 y_2^2 r_2 + y_1^3 y_4 r_1 + y_1^3 y_3 r_2 + y_1^3 y_2 r_2 + y_1^3 y_2 r_1$

105. $t_4 t_8 = y_1^3 q_6 + y_1^3 q_4 + y_1^3 q_2 + y_1^3 q_1 + y_1^2 y_3 y_4 r_2 + y_1^2 y_3 y_4 r_1 + y_1^2 y_2 y_4 r_1 + y_1^2 y_2^2 r_2 + y_1^2 y_2^2 r_1 + y_1^3 y_4 r_2 + y_1^3 y_4 r_1 + y_1^3 y_3 r_1 + y_1^4 r_2 + y_1^4 r_1$

106. $t_4 t_7 = y_1^2 y_2 q_4 + y_1^2 y_2 q_3 + y_1^3 q_5 + y_1^3 q_3 + y_1^2 y_2 y_3 r_2 + y_1^2 y_2^2 r_1 + y_1^3 y_4 r_2 + y_1^3 y_3 r_2 + y_1^4 r_2$

107. $t_4 t_6 = y_1^2 y_2 q_4 + y_1^2 y_2 q_3 + y_1^3 q_6 + y_1^3 q_4 + y_1^3 q_3 + y_1^3 q_2 + y_1^3 q_1 + y_1^2 y_3 y_4 r_2 + y_1^2 y_3 y_4 r_1 + y_1^2 y_2 y_3 r_1 + y_1^3 y_4 r_1 + y_1^3 y_3 r_2 + y_1^3 y_2 r_2$

108. $t_4 t_5 = y_1^3 q_6 + y_1^3 q_5 + y_1^3 q_4 + y_1^3 q_3 + y_1^3 q_1 + y_1^2 y_3 y_4 r_2 + y_1^2 y_2 y_4 r_2 + y_1^2 y_2 y_4 r_1 + y_1^2 y_2^2 r_2 + y_1^3 y_2 r_2 + y_1^3 y_2 r_1 + y_1^4 r_2$

109. $t_4^2 = y_1^2 y_2 y_4 r_2 + y_1^2 y_2 y_4 r_1 + y_1^2 y_2 y_3 r_1 + y_1^2 y_2^2 r_1 + y_1^3 y_4 r_1 + y_1^3 y_3 r_2 + y_1^3 y_2 r_2 + y_1^3 y_2 r_1 + y_1^4 r_2$

110. $t_3 t_8 = y_1^3 q_5 + y_1^3 q_4 + y_1^3 q_3 + y_1^2 y_3 y_4 r_1 + y_1^2 y_2^2 r_2 + y_1^2 y_2^2 r_1 + y_1^3 y_2 r_2 + y_1^4 r_1$

111. $t_3 t_7 = y_1^2 y_2 q_4 + y_1^2 y_2 q_3 + y_1^3 q_3 + y_1^3 q_2 + y_1^3 q_1 + y_1^2 y_2 y_4 r_2 + y_1^2 y_2 y_4 r_1 + y_1^2 y_2 y_3 r_2 + y_1^2 y_2^2 r_1 + y_1^4 r_1$

112. $t_3 t_6 = y_1^2 y_2 q_4 + y_1^3 q_6 + y_1^3 q_5 + y_1^3 q_4 + y_1^3 q_2 + y_1^2 y_3 y_4 r_1 + y_1^2 y_2 y_3 r_2 + y_1^2 y_2^2 r_1 + y_1^3 y_4 r_2 + y_1^3 y_3 r_2 + y_1^3 y_3 r_1 + y_1^3 y_2 r_2 + y_1^4 r_2$

113. $t_3 t_5 = y_1^2 y_2 q_3 + y_1^3 q_6 + y_1^3 q_5 + y_1^3 q_4 + y_1^3 q_3 + y_1^3 q_1 + y_1^2 y_3 y_4 r_2 + y_1^2 y_3 y_4 r_1 + y_1^2 y_2 y_4 r_2 + y_1^2 y_2 y_4 r_1 + y_1^2 y_2^2 r_2 + y_1^2 y_2^2 r_1 + y_1^3 y_4 r_2 + y_1^3 y_4 r_1 + y_1^3 y_2 r_2 + y_1^4 r_1$

114. $t_3 t_4 = y_1^3 q_6 + y_1^3 q_5 + y_1^3 q_3 + y_1^3 q_1 + y_1^2 y_3 y_4 r_2 + y_1^2 y_3 y_4 r_1 + y_1^2 y_2 y_4 r_2 + y_1^2 y_2 y_3 r_2 + y_1^2 y_2^2 r_2 + y_1^2 y_2^2 r_1 + y_1^3 y_4 r_2 + y_1^3 y_3 r_1$

115. $t_3^2 = y_1^2 y_2 y_4 r_1 + y_1^2 y_2 y_3 r_2 + y_1^3 y_4 r_2 + y_1^3 y_3 r_2 + y_1^3 y_3 r_1 + y_1^3 y_2 r_1 + y_1^4 r_2$

116. $t_2 t_8 = y_1^3 q_4 + y_1^3 q_2 + y_1^2 y_2 y_4 r_1 + y_1^2 y_2 y_3 r_1 + y_1^2 y_2^2 r_1 + y_1^3 y_4 r_2 + y_1^3 y_3 r_2 + y_1^3 y_3 r_1 + y_1^3 y_2 r_2 + y_1^4 r_2 + y_1^4 r_1$

117. $t_2 t_7 = y_1^3 q_3 + y_1^3 q_1 + y_1^2 y_3 y_4 r_2 + y_1^2 y_3 y_4 r_1 + y_1^2 y_2 y_4 r_2 + y_1^2 y_2 y_4 r_1 + y_1^2 y_2^2 r_1 + y_1^3 y_3 r_1$

118. $t_2 t_6 = y_1^2 y_2 q_3 + y_1^3 q_2 + y_1^3 q_1 + y_1^2 y_3 y_4 r_1 + y_1^2 y_2^2 r_2 + y_1^2 y_2^2 r_1 + y_1^3 y_4 r_2 + y_1^3 y_3 r_2 + y_1^3 y_2 r_1 + y_1^4 r_2$

119. $t_2 t_5 = y_1^2 y_2 q_4 + y_1^3 q_6 + y_1^3 q_3 + y_1^2 y_2 y_3 r_2 + y_1^2 y_2 y_3 r_1 + y_1^2 y_2^2 r_1 + y_1^3 y_4 r_2 + y_1^3 y_4 r_1 + y_1^3 y_3 r_2 + y_1^3 y_3 r_1 + y_1^3 y_2 r_2 + y_1^4 r_2 + y_1^4 r_1$

120. $t_2t_4 = y_1^2y_2q_4 + y_1^2y_2q_3 + y_1^3q_5 + y_1^3q_4 + y_1^3q_1 + y_1^2y_2y_4r_2 + y_1^2y_2y_4r_1 + y_1^2y_2y_3r_1 + y_1^3y_3r_2 + y_1^3y_2r_2 + y_1^3y_2r_1$

121. $t_2t_3 = y_1^2y_2q_3 + y_1^3q_6 + y_1^3q_2 + y_1^2y_3y_4r_1 + y_1^2y_2y_4r_2 + y_1^2y_2y_4r_1 + y_1^2y_2y_3r_1 + y_1^2y_2^2r_1 + y_1^3y_4r_1 + y_1^3y_3r_2 + y_1^3y_3r_1 + y_1^3y_2r_2 + y_1^4r_1$

122. $t_2^2 = y_1^2y_2^2r_2 + y_1^2y_2^2r_1 + y_1^4r_2$

123. $t_1t_8 = y_1^3q_6 + y_1^3q_5 + y_1^3q_3 + y_1^2y_3y_4r_2 + y_1^2y_2y_3r_1 + y_1^3y_4r_2 + y_1^3y_4r_1 + y_1^3y_2r_2 + y_1^4r_1$

124. $t_1t_7 = y_1^2y_2q_4 + y_1^2y_2q_3 + y_1^3q_5 + y_1^3q_4 + y_1^3q_3 + y_1^3q_1 + y_1^2y_3y_4r_1 + y_1^2y_2y_4r_2 + y_1^2y_2y_4r_1 + y_1^2y_2y_3r_2 + y_1^2y_2y_3r_1 + y_1^2y_2^2r_1 + y_1^3y_4r_2 + y_1^3y_2r_1$

125. $t_1t_6 = y_1^2y_2q_3 + y_1^3q_3 + y_1^3q_2 + y_1^2y_3y_4r_1 + y_1^2y_2y_4r_2 + y_1^3y_3r_1 + y_1^3y_2r_2 + y_1^3y_2r_1 + y_1^4r_2$

126. $t_1t_5 = y_1^2y_2q_4 + y_1^2y_2q_3 + y_1^3q_2 + y_1^3q_1 + y_1^2y_3y_4r_1 + y_1^2y_2y_3r_2 + y_1^2y_2y_3r_1 + y_1^2y_2^2r_2 + y_1^2y_2^2r_1 + y_1^3y_4r_1 + y_1^3y_2r_2 + y_1^4r_1$

127. $t_1t_4 = y_1^2y_2q_3 + y_1^3q_4 + y_1^3q_3 + y_1^3q_1 + y_1^2y_3y_4r_1 + y_1^2y_2y_4r_2 + y_1^2y_2y_4r_1 + y_1^2y_2y_3r_2 + y_1^2y_2y_3r_1 + y_1^2y_2^2r_2 + y_1^2y_2^2r_1 + y_1^3y_4r_1 + y_1^3y_2r_2 + y_1^4r_2 + y_1^4r_1$

128. $t_1t_3 = y_1^2y_2q_3 + y_1^3q_6 + y_1^3q_5 + y_1^3q_4 + y_1^3q_3 + y_1^3q_1 + y_1^2y_2y_4r_2 + y_1^2y_2y_3r_1 + y_1^2y_2^2r_2 + y_1^3y_3r_2 + y_1^3y_2r_2 + y_1^4r_1$

129. $t_1t_2 = y_1^3q_4 + y_1^3q_3 + y_1^3q_2 + y_1^2y_2y_3r_1 + y_1^3y_4r_1 + y_1^3y_2r_1 + y_1^4r_1$

130. $t_1^2 = y_1^2y_2^2r_1 + y_1^4r_2 + y_1^4r_1$

131. $y_4o_2 = y_1^2y_2q_4 + y_1^2y_2q_3 + y_1^3q_6 + y_1^3q_4 + y_1^3q_2 + y_1^3q_1 + y_1^2y_3y_4r_1 + y_1^2y_2y_4r_2 + y_1^2y_2y_3r_1 + y_1^2y_2^2r_2 + y_1^3y_3r_2 + y_1^3y_3r_1$

132. $y_4o_1 = y_1^2y_2q_4 + y_1^2y_2q_3 + y_1^3q_6 + y_1^2y_2y_4r_2 + y_1^2y_2y_4r_1 + y_1^2y_2^2r_2 + y_1^3y_4r_2 + y_1^3y_2r_2 + y_1^4r_2 + y_1^4r_1$

133. $y_3o_2 = y_1^2y_2q_4 + y_1^2y_2q_3 + y_1^3q_4 + y_1^3q_3 + y_1^2y_3y_4r_1 + y_1^2y_2y_3r_2 + y_1^2y_2^2r_2 + y_1^2y_2^2r_1 + y_1^3y_4r_1 + y_1^4r_2 + y_1^4r_1$

134. $y_3o_1 = y_1^3q_3 + y_1^3q_1 + y_1^2y_3y_4r_1 + y_1^2y_2y_3r_2 + y_1^3y_4r_2 + y_1^3y_4r_1 + y_1^3y_3r_2 + y_1^3y_2r_1 + y_1^4r_1$

135. $y_2o_2 = y_1^3q_6 + y_1^3q_4 + y_1^3q_3 + y_1^3q_2 + y_1^3q_1 + y_1^2y_2y_4r_1 + y_1^2y_2^2r_2 + y_1^2y_2^2r_1 + y_1^3y_4r_2 + y_1^3y_3r_2 + y_1^3y_2r_2 + y_1^3y_2r_1 + y_1^4r_1$

136. $y_2o_1 = y_1^2y_2q_4 + y_1^2y_2q_3 + y_1^3q_6 + y_1^3q_3 + y_1^3q_1 + y_1^2y_3y_4r_1 + y_1^2y_2y_4r_1 + y_1^3y_3r_2 + y_1^3y_2r_2 + y_1^3y_2r_1 + y_1^4r_2 + y_1^4r_1$

137. $y_1o_2 = y_1^2y_2q_4 + y_1^2y_2q_3 + y_1^3q_5 + y_1^3q_2 + y_1^2y_3y_4r_1 + y_1^2y_2y_4r_1 + y_1^2y_2y_3r_1 + y_1^2y_2^2r_2 + y_1^2y_2^2r_1 + y_1^3y_4r_2 + y_1^3y_3r_2 + y_1^3y_3r_1$

138. $y_1o_1 = y_1^2y_2q_4 + y_1^3q_4 + y_1^3q_3 + y_1^3q_2 + y_1^2y_2y_4r_2 + y_1^2y_2^2r_2 + y_1^3y_3r_1 + y_1^3y_2r_2 + y_1^4r_2$

139. $v_4q_6 = y_2v_4r_2 + y_2v_3r_1 + y_2v_2r_2 + y_2v_2r_1 + y_1v_4r_2 + y_1v_4r_1 + y_1v_3r_2 + y_1v_2r_2 + y_1v_2r_1 + y_1v_1r_2 + y_1v_1r_1 + y_1^4q_4 + y_1^4q_3 + y_1^3y_3y_4r_2 + y_1^3y_2y_4r_2 + y_1^3y_2y_4r_1 + y_1^4y_4r_2 + y_1^4y_4r_1 + y_1^4y_2r_1$

140. $v_4q_5 = y_2v_3r_2 + y_2v_2r_2 + y_2v_2r_1 + y_1v_3r_1 + y_1v_2r_2 + y_1v_2r_1 + y_1v_1r_2 + y_1v_1r_1 + y_1^4q_1 + y_1^3y_3y_4r_1 + y_1^3y_2y_4r_2 + y_1^3y_2y_4r_1 + y_1^4y_4r_2 + y_1^4y_3r_2 + y_1^4y_2r_1 + y_1^5r_2 + y_1^5r_1$

141. $v_4q_4 = y_2v_4r_1 + y_2v_2r_2 + y_2v_2r_1 + y_1v_4r_2 + y_1v_4r_1 + y_1v_3r_2 + y_1v_3r_1 + y_1v_2r_2 + y_1v_2r_1 + y_1v_1r_2 + y_1v_1r_1 + y_1^4q_4 + y_1^3y_3y_4r_2 + y_1^3y_3y_4r_1 + y_1^3y_2y_4r_2 + y_1^3y_2y_4r_1 + y_1^4y_2r_1$

142. $v_4q_3 = y_2v_4r_2 + y_2v_2r_2 + y_2v_2r_1 + y_1v_3r_2 + y_1v_3r_1 + y_1v_2r_2 + y_1v_2r_1 + y_1v_1r_2 + y_1v_1r_1 + y_1^4q_3 + y_1^4q_2 + y_1^4q_1 + y_1^3y_3y_4r_2 + y_1^3y_3y_4r_1 + y_1^3y_2y_4r_2 + y_1^3y_2y_4r_1 + y_1^4y_4r_1 + y_1^4y_3r_2 + y_1^4y_2r_2 + y_1^5r_2 + y_1^5r_1$

143. $v_4q_2 = y_1^4q_4 + y_1^4q_3 + y_1^4q_2 + y_1^4q_1 + y_1^4y_4r_1 + y_1^4y_3r_2 + y_1^4y_2r_2 + y_1^4y_2r_1 + y_1^5r_2 + y_1^5r_1$

144. $v_4q_1 = y_2v_4r_2 + y_2v_4r_1 + y_2v_3r_2 + y_2v_2r_1 + y_1v_3r_2 + y_1v_3r_1 + y_1v_2r_1 + y_1v_1r_1 + y_1^4q_2 + y_1^4q_1 + y_1^3y_3y_4r_2 + y_1^3y_3y_4r_1 + y_1^3y_2y_4r_1 + y_1^4y_3r_2 + y_1^4y_2r_2 + y_1^4y_2r_1 + y_1^5r_1$

145. $v_3q_6 = y_2v_4r_1 + y_2v_2r_2 + y_1v_4r_2 + y_1v_2r_2 + y_1v_2r_1 + y_1^4q_4 + y_1^4q_3 + y_1^4q_1 + y_1^4y_2r_2$

146. $v_3q_5 = y_2v_4r_2 + y_2v_3r_1 + y_2v_2r_1 + y_1v_4r_1 + y_1v_3r_2 + y_1v_3r_1 + y_1v_2r_2 + y_1v_2r_1 + y_1^4q_3 + y_1^4q_1 + y_1^4y_3r_2 + y_1^4y_3r_1 + y_1^4y_2r_2 + y_1^4y_2r_1 + y_1^5r_2 + y_1^5r_1$

147. $v_3q_4 = y_2v_3r_2 + y_2v_2r_2 + y_2v_2r_1 + y_1v_4r_2 + y_1v_4r_1 + y_1v_2r_2 + y_1v_2r_1 + y_1^4q_4 + y_1^4q_1 + y_1^4y_4r_2 + y_1^4y_3r_1 + y_1^4y_2r_2$

148. $v_3q_3 = y_2v_3r_1 + y_2v_2r_2 + y_2v_2r_1 + y_1v_4r_2 + y_1v_4r_1 + y_1v_3r_2 + y_1v_3r_1 + y_1v_2r_2 + y_1v_2r_1 + y_1^4q_3 + y_1^4q_1 + y_1^4y_4r_2 + y_1^4y_3r_1 + y_1^4y_2r_1 + y_1^5r_2 + y_1^5r_1$

149. $v_3q_2 = y_1^4q_4 + y_1^4q_3 + y_1^4q_2 + y_1^4y_2r_2 + y_1^4y_2r_1 + y_1^5r_2 + y_1^5r_1$

150. $v_3q_1 = y_2v_4r_2 + y_2v_2r_2 + y_2v_2r_1 + y_1v_4r_2 + y_1v_4r_1 + y_1v_3r_1 + y_1v_2r_1 + y_1^4q_3 + y_1^4y_4r_1 + y_1^4y_3r_2 + y_1^4y_3r_1 + y_1^4y_2r_2 + y_1^4y_2r_1 + y_1^5r_1$

151. $v_2q_6 = y_2v_4r_2 + y_2v_4r_1 + y_2v_2r_1 + y_1v_4r_1 + y_1v_2r_2 + y_1v_2r_1 + y_1v_1r_2 + y_1^4q_3 + y_1^3y_3y_4r_2 + y_1^3y_3y_4r_1 + y_1^3y_2y_4r_2 + y_1^3y_2y_4r_1 + y_1^4y_4r_2 + y_1^4y_4r_1 + y_1^4y_3r_1 + y_1^4y_2r_2 + y_1^4y_2r_1 + y_1^5r_2$

152. $v_2q_5 = y_2v_4r_2 + y_2v_4r_1 + y_2v_2r_2 + y_2v_2r_1 + y_1v_4r_2 + y_1v_1r_1 + y_1^4q_3 + y_1^4q_1 + y_1^3y_3y_4r_2 + y_1^3y_3y_4r_1 + y_1^3y_2y_4r_2 + y_1^3y_2y_4r_1 + y_1^4y_4r_2 + y_1^5r_1$

153. $v_2q_4 = y_2v_4r_2 + y_2v_4r_1 + y_2v_2r_2 + y_1v_2r_2 + y_1v_2r_1 + y_1v_1r_2 + y_1v_1r_1 + y_1^4q_4 + y_1^4q_1 + y_1^3y_3y_4r_2 + y_1^3y_3y_4r_1 + y_1^3y_2y_4r_2 + y_1^3y_2y_4r_1 + y_1^4y_4r_2 + y_1^4y_3r_2 + y_1^4y_2r_1$

154. $v_2q_3 = y_2v_4r_2 + y_2v_4r_1 + y_2v_2r_1 + y_1v_1r_2 + y_1v_1r_1 + y_1^4q_4 + y_1^4q_3 + y_1^4q_1 + y_1^3y_3y_4r_2 + y_1^3y_3y_4r_1 + y_1^3y_2y_4r_2 + y_1^3y_2y_4r_1 + y_1^4y_4r_1 + y_1^4y_3r_2$

155. $v_2q_2 = y_1^4q_2 + y_1^4q_1 + y_1^4y_4r_1 + y_1^4y_3r_2$

156. $v_2q_1 = y_2v_4r_1 + y_2v_2r_2 + y_1v_4r_2 + y_1v_1r_2 + y_1v_1r_1 + y_1^4q_4 + y_1^4q_2 + y_1^3y_3y_4r_1 + y_1^3y_2y_4r_1 + y_1^4y_4r_2 + y_1^4y_4r_1 + y_1^4y_3r_2 + y_1^4y_2r_1$

157. $v_1q_6 = y_2v_3r_2 + y_2v_3r_1 + y_2v_2r_1 + y_1v_3r_1 + y_1v_2r_1 + y_1v_1r_1 + y_1^4q_4 + y_1^4q_3 + y_1^3y_3y_4r_2 + y_1^3y_3y_4r_1 + y_1^3y_2y_4r_1 + y_1^4y_3r_2 + y_1^5r_1$

158. $v_1q_5 = y_2v_3r_2 + y_2v_3r_1 + y_2v_2r_2 + y_1v_3r_2 + y_1v_2r_2 + y_1v_2r_1 + y_1v_1r_1 + y_1^4q_4 + y_1^4q_3 + y_1^4q_2 + y_1^3y_3y_4r_2 + y_1^3y_3y_4r_1 + y_1^3y_2y_4r_2 + y_1^4y_2r_1 + y_1^5r_2$

159. $v_1q_4 = y_2v_3r_2 + y_2v_3r_1 + y_1v_2r_2 + y_1^4q_1 + y_1^3y_3y_4r_2 + y_1^3y_3y_4r_1 + y_1^4y_4r_1 + y_1^4y_3r_2 + y_1^4y_3r_1 + y_1^4y_2r_1 + y_1^5r_2 + y_1^5r_1$

160. $v_1q_3 = y_2v_3r_2 + y_2v_3r_1 + y_1v_2r_1 + y_1v_1r_2 + y_1v_1r_1 + y_1^4q_4 + y_1^4q_3 + y_1^4q_1 + y_1^3y_3y_4r_2 + y_1^3y_3y_4r_1 + y_1^4y_4r_1 + y_1^4y_2r_1$

161. $v_1q_2 = y_1^4q_4 + y_1^4y_4r_2 + y_1^4y_4r_1 + y_1^4y_2r_2 + y_1^5r_2 + y_1^5r_1$

162. $v_1q_1 = y_2v_3r_1 + y_2v_2r_2 + y_1v_3r_2 + y_1v_2r_2 + y_1v_1r_2 + y_1v_1r_1 + y_1^4q_3 + y_1^4q_2 + y_1^3y_3y_4r_1 + y_1^3y_2y_4r_2 + y_1^4y_4r_2 + y_1^4y_4r_1 + y_1^4y_3r_2 + y_1^4y_3r_1 + y_1^5r_2 + y_1^5r_1$

163. $t_8q_6 = y_2t_8r_2 + y_2t_7r_1 + y_2t_4r_2 + y_2t_3r_1 + y_1t_8r_2 + y_1t_7r_1 + y_1t_6r_2 + y_1t_5r_1 + y_1t_4r_1 + y_1t_3r_2 + y_1t_1r_2 + y_1y_2^2v_4r_1 + y_1^2y_2v_4r_2 + y_1^2y_2v_4r_1 + y_1^2y_2v_3r_2 + y_1^2y_2v_2r_2 + y_1^2y_2v_2r_1 + y_1^3v_4r_2 + y_1^3v_3r_1 + y_1^3v_2r_2 + y_1^3v_2r_1 + y_1^3v_1r_2 + y_1^3v_1r_1$

164. $t_8q_5 = y_2t_7r_2+y_2t_4r_1+y_2t_3r_2+y_1t_8r_2+y_1t_7r_2+y_1t_6r_1+y_1t_5r_2+y_1t_4r_2+$
$y_1t_3r_1+y_1t_1r_1+y_1y_2^2v_4r_2+y_1^2y_2v_4r_2+y_1^2y_2v_3r_2+y_1^3v_4r_1+y_1^3v_3r_2+y_1^3v_2r_2+$
$y_1^3v_1r_2+y_1^3v_1r_1$

165. $t_8q_4 = y_2t_8r_1+y_2t_4r_2+y_2t_4r_1+y_1t_8r_2+y_1t_8r_1+y_1t_6r_2+y_1t_6r_1+y_1t_3r_2+$
$y_1t_3r_1+y_1t_1r_2+y_1t_1r_1+y_1^2y_2v_3r_1+y_1^2y_2v_2r_1+y_1^3v_3r_2+y_1^3v_3r_1+y_1^3v_2r_1+$
$y_1^3v_1r_2+y_1^3v_1r_1$

166. $t_8q_3 = y_2t_8r_2+y_2t_4r_2+y_2t_4r_1+y_1t_6r_2+y_1t_6r_1+y_1t_3r_2+y_1t_3r_1+y_1t_1r_2+$
$y_1t_1r_1+y_1y_2^2v_4r_1+y_1^2y_2v_3r_2+y_1^2y_2v_3r_1+y_1^2y_2v_2r_2+y_1^3v_4r_1+y_1^3v_3r_2+$
$y_1^3v_3r_1+y_1^3v_2r_2+y_1^3v_1r_2+y_1^3v_1r_1$

167. $t_8q_2 = y_1y_2^2v_4r_2+y_1y_2^2v_4r_1+y_1^2y_2v_4r_2+y_1^2y_2v_4r_1+y_1^2y_2v_3r_2+y_1^2y_2v_3r_1+$
$y_1^3v_4r_1+y_1^3v_3r_2+y_1^3v_2r_2$

168. $t_8q_1 = y_2t_8r_2 + y_2t_8r_1 + y_2t_7r_2 + y_2t_4r_2 + y_2t_4r_1 + y_2t_3r_2 + y_1t_8r_2 +$
$y_1t_7r_2 + y_1t_6r_2 + y_1t_6r_1 + y_1t_5r_2 + y_1t_4r_2 + y_1t_3r_2 + y_1t_3r_1 + y_1t_1r_2 +$
$y_1t_1r_1 + y_1y_2^2v_4r_1 + y_1^2y_2v_4r_2 + y_1^2y_2v_3r_2 + y_1^2y_2v_2r_1 + y_1^3v_1r_2 + y_1^3v_1r_1$

169. $t_7q_6 = y_2t_8r_2+y_2t_3r_2+y_1t_7r_2+y_1t_6r_2+y_1t_6r_1+y_1t_5r_1+y_1t_4r_1+y_1t_3r_1+$
$y_1t_2r_1+y_1^2y_2v_2r_2+y_1^2y_2v_2r_1+y_1^3v_4r_2+y_1^3v_4r_1+y_1^3v_3r_2+y_1^3v_3r_1+y_1^3v_2r_2+$
$y_1^3v_2r_1 + y_1^3v_1r_1$

170. $t_7q_5 = y_2t_8r_1+y_2t_7r_1+y_2t_3r_1+y_1t_7r_2+y_1t_6r_2+y_1t_6r_1+y_1t_5r_2+y_1t_4r_2+$
$y_1t_3r_2 + y_1t_2r_2 + y_1y_2^2v_4r_2 + y_1^2y_2v_4r_2 + y_1^2y_2v_4r_1 + y_1^2y_2v_3r_2 + y_1^2y_2v_3r_1 +$
$y_1^2y_2v_2r_1 + y_1^3v_4r_2 + y_1^3v_3r_1 + y_1^3v_2r_1$

171. $t_7q_4 = y_2t_8r_2+y_2t_8r_1+y_2t_7r_2+y_2t_3r_2+y_2t_3r_1+y_1t_7r_2+y_1t_7r_1+y_1t_6r_2+$
$y_1t_6r_1 + y_1y_2^2v_4r_2 + y_1^2y_2v_3r_2 + y_1^2y_2v_3r_1 + y_1^3v_3r_2 + y_1^3v_3r_1 + y_1^3v_2r_2$

172. $t_7q_3 = y_2t_8r_2 + y_2t_8r_1 + y_2t_7r_1 + y_2t_3r_2 + y_2t_3r_1 + y_1t_6r_2 + y_1t_6r_1 +$
$y_1^2y_2v_3r_2 + y_1^3v_3r_2 + y_1^3v_3r_1 + y_1^3v_1r_1$

173. $t_7q_2 = y_1y_2^2v_4r_2 + y_1^2y_2v_4r_2 + y_1^2y_2v_4r_1 + y_1^2y_2v_3r_2 + y_1^3v_4r_2 + y_1^3v_4r_1 +$
$y_1^3v_3r_2 + y_1^3v_1r_1$

174. $t_7q_1 = y_2t_8r_2+y_2t_8r_1+y_2t_3r_2+y_2t_3r_1+y_1t_7r_2+y_1t_6r_1+y_1t_5r_2+y_1t_4r_2+$
$y_1t_3r_2 + y_1t_2r_2 + y_1^2y_2v_4r_2 + y_1^2y_2v_3r_1 + y_1^3v_2r_1 + y_1^3v_1r_2$

175. $t_6q_6 = y_2t_4r_2+y_2t_4r_1+y_2t_3r_1+y_1t_8r_1+y_1t_7r_2+y_1t_7r_1+y_1t_6r_2+y_1t_5r_2+$
$y_1t_3r_2+y_1t_2r_1+y_1t_1r_2+y_1t_1r_1+y_1^2y_2v_3r_1+y_1^2y_2v_2r_2+y_1^3v_3r_2+y_1^3v_2r_2+$
$y_1^3v_2r_1 + y_1^3v_1r_1$

176. $t_6q_5 = y_2t_4r_2 + y_2t_3r_2 + y_2t_3r_1 + y_1t_8r_2 + y_1t_7r_2 + y_1t_7r_1 + y_1t_6r_2 +$
$y_1t_6r_1+y_1t_5r_1+y_1t_4r_1+y_1t_2r_2+y_1t_1r_2+y_1^2y_2v_4r_1+y_1^2y_2v_3r_1+y_1^2y_2v_2r_2+$
$y_1^2y_2v_2r_1 + y_1^3v_4r_2 + y_1^3v_2r_1 + y_1^3v_1r_2$

177. $t_6q_4 = y_2t_4r_1+y_2t_3r_2+y_1t_7r_2+y_1t_7r_1+y_1t_6r_1+y_1t_5r_2+y_1t_5r_1+y_1t_4r_2+$
$y_1t_3r_1+y_1t_1r_1+y_1^2y_2v_4r_1+y_1^2y_2v_2r_1+y_1^3v_4r_2+y_1^3v_3r_2+y_1^3v_3r_1+y_1^3v_2r_1+$
$y_1^3v_1r_1$

178. $t_6q_3 = y_2t_4r_2+y_2t_3r_1+y_1t_7r_2+y_1t_7r_1+y_1t_6r_1+y_1t_5r_2+y_1t_5r_1+y_1t_4r_1+$
$y_1t_3r_2+y_1t_1r_2+y_1^2y_2v_4r_2+y_1^2y_2v_2r_2+y_1^2y_2v_2r_1+y_1^3v_3r_2+y_1^3v_2r_2+y_1^3v_1r_2+$
$y_1^3v_1r_1$

179. $t_6q_2 = y_1y_2^2v_4r_2+y_1^2y_2v_3r_2+y_1^2y_2v_3r_1+y_1^2y_2v_2r_2+y_1^3v_4r_2+y_1^3v_2r_1+y_1^3v_1r_2$

180. $t_6q_1 = y_2t_4r_1+y_2t_3r_2+y_1t_8r_2+y_1t_7r_1+y_1t_6r_2+y_1t_5r_2+y_1t_5r_1+y_1t_3r_2+$
$y_1t_3r_1 + y_1t_2r_2 + y_1t_1r_1 + y_1y_2^2v_4r_2 + y_1y_2^2v_4r_1 + y_1^2y_2v_4r_1 + y_1^2y_2v_3r_2 +$
$y_1^2y_2v_2r_2 + y_1^2y_2v_2r_1 + y_1^3v_4r_2 + y_1^3v_1r_2 + y_1^3v_1r_1$

181. $t_5q_6 = y_2t_8r_2+y_2t_8r_1+y_2t_4r_1+y_2t_3r_1+y_1t_8r_2+y_1t_8r_1+y_1t_7r_1+y_1t_5r_2+$
$y_1t_5r_1+y_1t_4r_2+y_1t_4r_1+y_1t_3r_1+y_1t_2r_1+y_1t_1r_2+y_1y_2^2v_4r_2+y_1y_2^2v_4r_1+$
$y_1^2y_2v_4r_1+y_1^2y_2v_3r_2+y_1^2y_2v_3r_1+y_1^2y_2v_2r_1+y_1^3v_4r_2+y_1^3v_3r_1+y_1^3v_2r_2+$
$y_1^3v_1r_2$

182. $t_5q_5 = y_2t_8r_2+y_2t_8r_1+y_2t_4r_2+y_2t_4r_1+y_2t_3r_2+y_1t_8r_2+y_1t_8r_1+y_1t_7r_2+$
$y_1t_6r_1+y_1t_5r_1+y_1t_4r_2+y_1t_4r_1+y_1t_3r_2+y_1t_3r_1+y_1t_2r_2+y_1t_1r_1+$
$y_1^2y_2v_4r_2+y_1^2y_2v_4r_1+y_1^2y_2v_3r_1+y_1^2y_2v_2r_1+y_1^3v_3r_2+y_1^3v_2r_2+y_1^3v_1r_2$

183. $t_5q_4 = y_2t_8r_2+y_2t_8r_1+y_2t_4r_2+y_1t_8r_2+y_1t_8r_1+y_1t_6r_2+y_1t_5r_1+$
$y_1t_4r_2+y_1t_4r_1+y_1t_3r_2+y_1t_1r_2+y_1t_1r_1+y_1y_2^2v_4r_2+y_1^2y_2v_4r_1+y_1^2y_2v_3r_2+$
$y_1^2y_2v_2r_2+y_1^2y_2v_2r_1+y_1^3v_4r_2+y_1^3v_4r_1+y_1^3v_3r_2+y_1^3v_2r_2+y_1^3v_1r_2$

184. $t_5q_3 = y_2t_8r_2+y_2t_8r_1+y_2t_4r_1+y_1t_8r_2+y_1t_8r_1+y_1t_6r_1+y_1t_5r_1+y_1t_4r_2+$
$y_1t_4r_1+y_1t_3r_1+y_1t_1r_2+y_1t_1r_1+y_1y_2^2v_4r_1+y_1^2y_2v_4r_2+y_1^2y_2v_2r_2+y_1^3v_4r_2+$
$y_1^3v_4r_1+y_1^3v_3r_1+y_1^3v_1r_1$

185. $t_5q_2 = y_1y_2^2v_4r_2+y_1^2y_2v_4r_2+y_1^2y_2v_3r_2+y_1^3v_4r_2+y_1^3v_4r_1+y_1^3v_3r_1+y_1^3v_2r_1$

186. $t_5q_1 = y_2t_8r_1+y_2t_4r_2+y_2t_3r_2+y_1t_8r_1+y_1t_7r_2+y_1t_4r_1+y_1t_3r_2+y_1t_2r_2+$
$y_1t_1r_2+y_1t_1r_1+y_1y_2^2v_4r_2+y_1y_2^2v_4r_1+y_1^2y_2v_3r_1+y_1^2y_2v_2r_1+y_1^3v_4r_1+$
$y_1^3v_3r_2+y_1^3v_3r_1+y_1^3v_1r_2$

187. $t_4q_6 = y_2t_8r_2+y_2t_8r_1+y_2t_7r_2+y_2t_7r_1+y_2t_4r_2+y_2t_3r_2+y_1t_8r_2+y_1t_8r_1+$
$y_1t_6r_2+y_1t_5r_2+y_1t_5r_1+y_1t_3r_2+y_1t_2r_2+y_1t_2r_1+y_1y_2^2v_4r_2+y_1^2y_2v_4r_1+$
$y_1^2y_2v_2r_2+y_1^3v_4r_1+y_1^3v_3r_2+y_1^3v_3r_1+y_1^3v_2r_1+y_1^3v_1r_2$

188. $t_4q_5 = y_2t_8r_2+y_2t_8r_1+y_2t_7r_2+y_2t_7r_1+y_2t_3r_1+y_1t_8r_2+y_1t_8r_1+y_1t_6r_1+$
$y_1t_5r_2+y_1t_5r_1+y_1t_4r_2+y_1t_4r_1+y_1t_3r_1+y_1t_2r_2+y_1t_2r_1+y_1y_2^2v_4r_2+$
$y_1^2y_2v_4r_1+y_1^2y_2v_3r_2+y_1^2y_2v_2r_2+y_1^2y_2v_2r_1+y_1^3v_4r_2+y_1^3v_4r_1+y_1^3v_3r_2+$
$y_1^3v_3r_1+y_1^3v_2r_2+y_1^3v_2r_1$

189. $t_4q_4 = y_2t_8r_2+y_2t_8r_1+y_2t_7r_2+y_2t_7r_1+y_2t_4r_1+y_2t_3r_2+y_2t_3r_1+y_1t_8r_2+$
$y_1t_8r_1+y_1t_6r_2+y_1t_6r_1+y_1t_5r_2+y_1t_5r_1+y_1t_3r_2+y_1t_3r_1+y_1t_2r_2+y_1t_2r_1+$
$y_1y_2^2v_4r_2+y_1^2y_2v_4r_2+y_1^2y_2v_4r_1+y_1^2y_2v_3r_2+y_1^2y_2v_2r_2+y_1^3v_4r_2+y_1^3v_4r_1+$
$y_1^3v_3r_2+y_1^3v_2r_2+y_1^3v_1r_2+y_1^3v_1r_1$

190. $t_4q_3 = y_2t_8r_2+y_2t_8r_1+y_2t_7r_2+y_2t_7r_1+y_2t_4r_2+y_2t_3r_2+y_2t_3r_1+y_1t_8r_2+$
$y_1t_8r_1+y_1t_6r_2+y_1t_6r_1+y_1t_5r_2+y_1t_5r_1+y_1t_4r_2+y_1t_4r_1+y_1t_3r_2+y_1t_3r_1+$
$y_1t_2r_2+y_1t_2r_1+y_1y_2^2v_4r_1+y_1^2y_2v_4r_2+y_1^2y_2v_2r_1+y_1^3v_4r_2+y_1^3v_4r_1+y_1^3v_3r_1+$
$y_1^3v_1r_2$

191. $t_4q_2 = y_1y_2^2v_4r_2+y_1^2y_2v_4r_2+y_1^2y_2v_3r_2+y_1^2y_2v_2r_2+y_1^2y_2v_2r_1+y_1^3v_4r_2+$
$y_1^3v_4r_1+y_1^3v_3r_1+y_1^3v_2r_1+y_1^3v_1r_2$

192. $t_4q_1 = y_2t_8r_1+y_2t_7r_1+y_2t_4r_2+y_2t_4r_1+y_2t_3r_2+y_2t_3r_1+y_1t_8r_1+y_1t_6r_2+$
$y_1t_6r_1+y_1t_5r_1+y_1t_4r_1+y_1t_3r_2+y_1t_3r_1+y_1t_2r_1+y_1y_2^2v_4r_1+y_1^2y_2v_4r_1+$
$y_1^2y_2v_3r_2+y_1^2y_2v_2r_2+y_1^3v_4r_1+y_1^3v_3r_2+y_1^3v_3r_1+y_1^3v_2r_2+y_1^3v_1r_1$

193. $t_3q_6 = y_2t_7r_2+y_2t_7r_1+y_2t_4r_1+y_2t_3r_2+y_2t_3r_1+y_1t_8r_2+y_1t_8r_1+y_1t_7r_2+$
$y_1t_7r_1+y_1t_6r_2+y_1t_6r_1+y_1t_5r_2+y_1t_3r_2+y_1t_3r_1+y_1t_2r_2+y_1y_2^2v_4r_2+$
$y_1^2y_2v_4r_2+y_1^2y_2v_3r_2+y_1^2y_2v_3r_1+y_1^2y_2v_2r_1+y_1^3v_4r_2+y_1^3v_3r_2+y_1^3v_3r_1+$
$y_1^3v_2r_2+y_1^3v_2r_1$

194. $t_3q_5 = y_2t_7r_2+y_2t_7r_1+y_2t_4r_2+y_2t_3r_2+y_1t_8r_2+y_1t_8r_1+y_1t_7r_2+y_1t_7r_1+$
$y_1t_6r_2+y_1t_6r_1+y_1t_5r_1+y_1t_2r_1+y_1^2y_2v_3r_2+y_1^3v_4r_2+y_1^3v_4r_1+y_1^3v_3r_2+$
$y_1^3v_1r_2+y_1^3v_1r_1$

195. $t_3q_4 = y_2t_7r_2 + y_2t_7r_1 + y_2t_3r_1 + y_1t_8r_2 + y_1t_8r_1 + y_1t_7r_2 + y_1t_7r_1 + y_1t_6r_2 +$
$y_1t_6r_1 + y_1t_5r_2 + y_1t_5r_1 + y_1t_3r_2 + y_1t_3r_1 + y_1t_2r_2 + y_1t_2r_1 + y_1^2y_2v_4r_1 +$
$y_1^2y_2v_3r_1 + y_1^3v_4r_2 + y_1^3v_4r_1 + y_1^3v_3r_1 + y_1^3v_2r_2 + y_1^3v_1r_2 + y_1^3v_1r_1$

196. $t_3q_3 = y_2t_7r_2 + y_2t_7r_1 + y_2t_3r_2 + y_1t_8r_2 + y_1t_8r_1 + y_1t_7r_2 + y_1t_7r_1 + y_1t_6r_2 +$
$y_1t_6r_1 + y_1t_5r_2 + y_1t_5r_1 + y_1t_2r_2 + y_1t_2r_1 + y_1y_2^2v_4r_1 + y_1^2y_2v_4r_2 + y_1^2y_2v_3r_2 +$
$y_1^3v_4r_2 + y_1^3v_4r_1 + y_1^3v_3r_2 + y_1^3v_3r_1 + y_1^3v_1r_2 + y_1^3v_1r_1$

197. $t_3q_2 = y_1^2y_2v_4r_2 + y_1^2y_2v_3r_2 + y_1^2y_2v_2r_2 + y_1^3v_4r_2 + y_1^3v_3r_1 + y_1^3v_2r_1 + y_1^3v_1r_1$

198. $t_3q_1 = y_2t_7r_1 + y_2t_4r_2 + y_2t_3r_1 + y_1t_8r_1 + y_1t_7r_1 + y_1t_6r_1 + y_1t_5r_2 + y_1t_5r_1 +$
$y_1t_2r_2 + y_1t_2r_1 + y_1y_2^2v_4r_2 + y_1y_2^2v_4r_1 + y_1^2y_2v_2r_1 + y_1^3v_4r_2 + y_1^3v_3r_2 + y_1^3v_3r_1$

199. $t_2q_6 = y_2t_4r_2 + y_2t_4r_1 + y_1t_6r_2 + y_1t_6r_1 + y_1t_5r_1 + y_1t_4r_2 + y_1t_4r_1 + y_1t_3r_2 +$
$y_1t_3r_1 + y_1t_2r_2 + y_1t_2r_1 + y_1y_2^2v_4r_2 + y_1y_2^2v_4r_1 + y_1^2y_2v_4r_2 + y_1^2y_2v_3r_2 +$
$y_1^2y_2v_3r_1 + y_1^2y_2v_2r_1 + y_1^3v_4r_2 + y_1^3v_3r_2 + y_1^3v_2r_1$

200. $t_2q_5 = y_2t_4r_2 + y_2t_4r_1 + y_1t_6r_2 + y_1t_6r_1 + y_1t_5r_2 + y_1t_5r_1 + y_1t_4r_2 + y_1t_3r_2 +$
$y_1t_3r_1 + y_1t_2r_1 + y_1t_1r_1 + y_1y_2^2v_4r_2 + y_1^2y_2v_2r_2 + y_1^2y_2v_2r_1 + y_1^3v_4r_1 + y_1^3v_3r_2 +$
$y_1^3v_3r_1 + y_1^3v_2r_2 + y_1^3v_2r_1 + y_1^3v_1r_1$

201. $t_2q_4 = y_2t_4r_2 + y_2t_4r_1 + y_1t_6r_2 + y_1t_6r_1 + y_1t_5r_2 + y_1t_4r_1 + y_1t_3r_2 + y_1t_3r_1 +$
$y_1t_2r_1 + y_1t_1r_2 + y_1y_2^2v_4r_2 + y_1^2y_2v_4r_2 + y_1^2y_2v_3r_2 + y_1^2y_2v_2r_2 + y_1^3v_4r_1 +$
$y_1^3v_3r_2 + y_1^3v_2r_1 + y_1^3v_1r_1$

202. $t_2q_3 = y_2t_4r_2 + y_2t_4r_1 + y_1t_6r_2 + y_1t_6r_1 + y_1t_5r_1 + y_1t_4r_2 + y_1t_3r_2 + y_1t_3r_1 +$
$y_1t_2r_1 + y_1t_1r_1 + y_1y_2^2v_4r_1 + y_1^2y_2v_4r_1 + y_1^2y_2v_3r_1 + y_1^2y_2v_2r_1 + y_1^3v_4r_2 +$
$y_1^3v_4r_1 + y_1^3v_2r_2 + y_1^3v_2r_1 + y_1^3v_1r_2 + y_1^3v_1r_1$

203. $t_2q_2 = y_1^2y_2v_3r_2 + y_1^2y_2v_2r_2 + y_1^3v_4r_2 + y_1^3v_2r_2 + y_1^3v_2r_1 + y_1^3v_1r_2 + y_1^3v_1r_1$

204. $t_2q_1 = y_2t_4r_1 + y_1t_6r_1 + y_1t_5r_2 + y_1t_4r_1 + y_1t_3r_1 + y_1y_2^2v_4r_2 + y_1^2y_2v_3r_2 +$
$y_1^2y_2v_3r_1 + y_1^2y_2v_2r_2 + y_1^2y_2v_2r_1 + y_1^3v_4r_2 + y_1^3v_4r_1 + y_1^3v_3r_1 + y_1^3v_2r_2 + y_1^3v_1r_2$

205. $t_1q_6 = y_2t_3r_2 + y_2t_3r_1 + y_1t_5r_2 + y_1t_4r_2 + y_1t_3r_1 + y_1t_2r_2 + y_1t_2r_1 + y_1t_1r_2 +$
$y_1t_1r_1 + y_1y_2^2v_4r_2 + y_1^2y_2v_3r_2 + y_1^2y_2v_3r_1 + y_1^2y_2v_2r_1 + y_1^3v_4r_1 + y_1^3v_3r_1 +$
$y_1^3v_2r_1 + y_1^3v_1r_2$

206. $t_1q_5 = y_2t_3r_2 + y_2t_3r_1 + y_1t_5r_1 + y_1t_4r_1 + y_1t_3r_2 + y_1t_2r_2 + y_1t_1r_1 + y_1y_2^2v_4r_1 +$
$y_1^2y_2v_3r_2 + y_1^2y_2v_2r_2 + y_1^2y_2v_2r_1 + y_1^3v_4r_2 + y_1^3v_4r_1 + y_1^3v_3r_1 + y_1^3v_2r_2$

207. $t_1q_4 = y_2t_3r_2 + y_2t_3r_1 + y_1t_5r_2 + y_1t_5r_1 + y_1t_4r_2 + y_1t_4r_1 + y_1t_2r_1 +$
$y_1t_1r_1 + y_1y_2^2v_4r_2 + y_1y_2^2v_4r_1 + y_1^2y_2v_4r_2 + y_1^2y_2v_4r_1 + y_1^2y_2v_3r_2 + y_1^2y_2v_3r_1 +$
$y_1^2y_2v_2r_2 + y_1^3v_3r_2 + y_1^3v_3r_1$

208. $t_1q_3 = y_2t_3r_2 + y_2t_3r_1 + y_1t_5r_2 + y_1t_5r_1 + y_1t_4r_2 + y_1t_4r_1 + y_1t_2r_2 + y_1t_1r_1 +$
$y_1y_2^2v_4r_2 + y_1y_2^2v_4r_1 + y_1^2y_2v_4r_1 + y_1^2y_2v_4r_1 + y_1^2y_2v_3r_2 + y_1^2y_2v_2r_1 + y_1^3v_4r_1 +$
$y_1^3v_3r_2$

209. $t_1q_2 = y_1^2y_2v_2r_2 + y_1^3v_1r_1$

210. $t_1q_1 = y_2t_3r_1 + y_1t_5r_2 + y_1t_5r_1 + y_1t_4r_2 + y_1t_4r_1 + y_1t_3r_2 + y_1t_2r_1 + y_1y_2^2v_4r_2 +$
$y_1y_2^2v_4r_1 + y_1^2y_2v_4r_2 + y_1^2y_2v_4r_1 + y_1^2y_2v_3r_2 + y_1^2y_2v_2r_1 + y_1^3v_4r_1 + y_1^3v_1r_1$

211. $v_4o_2 = y_1y_2^2v_4r_1 + y_1^2y_2v_4r_2 + y_1^2y_2v_2r_1 + y_1^3v_3r_1 + y_1^3v_1r_1$

212. $v_4o_1 = y_1y_2^2v_4r_1 + y_1^2y_2v_4r_2 + y_1^2y_2v_4r_1 + y_1^2y_2v_3r_1 + y_1^2y_2v_2r_1 + y_1^3v_3r_1 +$
$y_1^3v_1r_1$

213. $v_3o_2 = y_1y_2^2v_4r_1 + y_1^2y_2v_3r_2 + y_1^2y_2v_3r_1 + y_1^2y_2v_2r_1 + y_1^3v_4r_1 + y_1^3v_2r_1$

214. $v_3o_1 = y_1^2y_2v_3r_2 + y_1^3v_3r_1 + y_1^3v_2r_1 + y_1^3v_1r_1$

215. $v_2o_2 = y_1^2y_2v_4r_1 + y_1^2y_2v_2r_2 + y_1^3v_3r_1 + y_1^3v_2r_1 + y_1^3v_1r_1$

216. $v_2 o_1 = y_1 y_2^2 v_4 r_1 + y_1^2 y_2 v_2 r_2 + y_1^2 y_2 v_2 r_1 + y_1^3 v_4 r_1 + y_1^3 v_1 r_1$

217. $v_1 o_2 = y_1 y_2^2 v_4 r_1 + y_1^2 y_2 v_2 r_1 + y_1^3 v_2 r_2 + y_1^3 v_2 r_1$

218. $v_1 o_1 = y_1^2 y_2 v_4 r_1 + y_1^2 y_2 v_2 r_1 + y_1^3 v_3 r_1 + y_1^3 v_2 r_2 + y_1^3 v_2 r_1 + y_1^3 v_1 r_1$

219. $t_8 o_2 = y_1^3 t_8 r_2 + y_1^3 t_7 r_1 + y_1^3 t_6 r_1 + y_1^3 t_3 r_2 + y_1^3 t_3 r_1 + y_1^3 t_2 r_2 + y_1^3 t_2 r_1 + y_1^3 t_1 r_2 + y_1^3 t_1 r_1$

220. $t_8 o_1 = y_1^3 t_8 r_2 + y_1^3 t_7 r_1 + y_1^3 t_4 r_1 + y_1^3 t_3 r_2 + y_1^3 t_2 r_2 + y_1^3 t_2 r_1 + y_1^3 t_1 r_2 + y_1^3 t_1 r_1$

221. $t_7 o_2 = y_1^3 t_8 r_2 + y_1^3 t_7 r_1 + y_1^3 t_6 r_2 + y_1^3 t_6 r_1 + y_1^3 t_4 r_2 + y_1^3 t_4 r_1 + y_1^3 t_3 r_2 + y_1^3 t_3 r_1 + y_1^3 t_2 r_2$

222. $t_7 o_1 = y_1^3 t_8 r_2 + y_1^3 t_6 r_2 + y_1^3 t_4 r_2 + y_1^3 t_3 r_2 + y_1^3 t_3 r_1 + y_1^3 t_2 r_2 + y_1^3 t_1 r_1$

223. $t_6 o_2 = y_1^3 t_7 r_2 + y_1^3 t_4 r_2 + y_1^3 t_3 r_2 + y_1^3 t_2 r_2 + y_1^3 t_2 r_1$

224. $t_6 o_1 = y_1^3 t_8 r_1 + y_1^3 t_7 r_2 + y_1^3 t_7 r_1 + y_1^3 t_4 r_2 + y_1^3 t_4 r_1 + y_1^3 t_3 r_2 + y_1^3 t_3 r_1 + y_1^3 t_2 r_2 + y_1^3 t_1 r_1$

225. $t_5 o_2 = y_1^3 t_7 r_2 + y_1^3 t_7 r_1 + y_1^3 t_6 r_2 + y_1^3 t_6 r_1 + y_1^3 t_4 r_2 + y_1^3 t_3 r_2 + y_1^3 t_2 r_2 + y_1^3 t_2 r_1 + y_1^3 t_1 r_2$

226. $t_5 o_1 = y_1^3 t_7 r_2 + y_1^3 t_7 r_1 + y_1^3 t_6 r_2 + y_1^3 t_6 r_1 + y_1^3 t_4 r_2 + y_1^3 t_3 r_2 + y_1^3 t_2 r_2 + y_1^3 t_1 r_2$

227. $t_4 o_2 = y_1^3 t_8 r_1 + y_1^3 t_7 r_2 + y_1^3 t_6 r_1 + y_1^3 t_2 r_1 + y_1^3 t_1 r_1$

228. $t_4 o_1 = y_1^3 t_7 r_2 + y_1^3 t_7 r_1 + y_1^3 t_6 r_1 + y_1^3 t_3 r_1$

229. $t_3 o_2 = y_1^3 t_7 r_1 + y_1^3 t_6 r_2 + y_1^3 t_6 r_1 + y_1^3 t_3 r_1 + y_1^3 t_2 r_2 + y_1^3 t_2 r_1 + y_1^3 t_1 r_2$

230. $t_3 o_1 = y_1^3 t_8 r_1 + y_1^3 t_6 r_2 + y_1^3 t_6 r_1 + y_1^3 t_2 r_2 + y_1^3 t_2 r_1 + y_1^3 t_1 r_2$

231. $t_2 o_2 = y_1^3 t_8 r_1 + y_1^3 t_6 r_1 + y_1^3 t_4 r_1 + y_1^3 t_1 r_1$

232. $t_2 o_1 = y_1^3 t_8 r_1 + y_1^3 t_6 r_1 + y_1^3 t_3 r_1 + y_1^3 t_2 r_1$

233. $t_1 o_2 = y_1^3 t_8 r_1 + y_1^3 t_7 r_1 + y_1^3 t_6 r_1 + y_1^3 t_2 r_2 + y_1^3 t_2 r_1 + y_1^3 t_1 r_2$

234. $t_1 o_1 = y_1^3 t_7 r_1 + y_1^3 t_4 r_1 + y_1^3 t_2 r_2 + y_1^3 t_1 r_2$

235. $q_6^2 = y_2 y_4 r_2^2 + y_2 y_4 r_1^2 + y_2 y_3 r_2^2 + y_2^2 r_1^2 + y_1 y_4 r_2^2 + y_1 y_3 r_1^2 + y_1 y_2 r_2^2 + y_1 y_2 r_1^2$

236. $q_5 q_6 = y_2 r_1 q_4 + y_2 r_1 q_3 + y_1 r_2 q_5 + y_1 r_2 q_3 + y_1 r_2 q_1 + y_1 r_1 q_6 + y_1 r_1 q_4 + y_1 r_1 q_2 + y_1 r_1 q_1 + y_3 y_4 r_2^2 + y_3 y_4 r_1^2 + y_2 y_4 r_2^2 + y_2 y_4 r_1 r_2 + y_2 y_3 r_2^2 + y_2 y_3 r_1 r_2 + y_2^2 r_1^2 + y_1 y_4 r_1^2 + y_1 y_3 r_2^2 + y_1 y_3 r_1 r_2 + y_1 y_3 r_1^2 + y_1 y_2 r_2^2 + y_1 y_2 r_1 r_2 + y_1 y_2 r_1^2 + y_1^2 r_2^2 + y_1^2 r_1^2 + y_1^4 t_4 r_1 + y_1^4 t_3 r_2 + y_1^4 t_3 r_1 + y_1^4 t_2 r_2 + y_1^4 t_2 r_1 + y_1^4 t_1 r_1$

237. $q_5^2 = y_2 y_4 r_2^2 + y_2 y_4 r_1^2 + y_2 y_3 r_1^2 + y_2^2 r_2^2 + y_2^2 r_1^2 + y_1 y_4 r_1^2 + y_1 y_3 r_2^2 + y_1 y_2 r_2^2 + y_1 y_2 r_1^2 + y_1^2 r_2^2 + y_1^2 r_1^2$

238. $q_4 q_6 = y_2 r_2 q_3 + y_1 r_2 q_6 + y_1 r_2 q_4 + y_1 r_2 q_3 + y_1 r_1 q_6 + y_1 r_1 q_5 + y_1 r_1 q_4 + y_3 y_4 r_1 r_2 + y_3 y_4 r_1^2 + y_2 y_4 r_2^2 + y_2 y_4 r_1^2 + y_2 y_3 r_2^2 + y_2 y_3 r_1 r_2 + y_2 y_3 r_1^2 + y_2^2 r_1 r_2 + y_1 y_3 r_1 r_2 + y_1 y_2 r_2^2 + y_1 y_2 r_1 r_2 + y_1^4 t_3 r_2 + y_1^4 t_3 r_1 + y_1^4 t_2 r_1 + y_1^4 t_1 r_1$

239. $q_4 q_5 = y_2 r_1 q_3 + y_1 r_2 q_5 + y_1 r_2 q_4 + y_1 r_2 q_3 + y_1 r_2 q_1 + y_1 r_1 q_4 + y_1 r_1 q_2 + y_3 y_4 r_2^2 + y_3 y_4 r_1 r_2 + y_2 y_4 r_1 r_2 + y_2 y_4 r_1^2 + y_2 y_3 r_1^2 + y_2^2 r_1 r_2 + y_2^2 r_1^2 + y_1 y_4 r_2^2 + y_1 y_4 r_1 r_2 + y_1 y_4 r_1^2 + y_1 y_3 r_2^2 + y_1 y_3 r_1 r_2 + y_1^4 t_4 r_2 + y_1^4 t_2 r_2 + y_1^4 t_1 r_1$

240. $q_4^2 = y_2 y_4 r_2^2 + y_2 y_4 r_1^2 + y_2 y_3 r_2^2 + y_2 y_3 r_1^2 + y_2^2 r_2^2 + y_1 y_4 r_2^2 + y_1 y_4 r_1^2 + y_1 y_2 r_2^2 + y_1 y_2 r_1^2$

241. $q_3 q_6 = y_2 r_2 q_4 + y_2 r_2 q_3 + y_2 r_1 q_4 + y_2 r_1 q_3 + y_1 r_1 q_6 + y_1 r_1 q_5 + y_1 r_1 q_3 + y_3 y_4 r_1 r_2 + y_3 y_4 r_1^2 + y_2 y_4 r_2^2 + y_2 y_4 r_1 r_2 + y_2 y_3 r_2^2 + y_2^2 r_2^2 + y_2^2 r_1^2 + y_1 y_4 r_1 r_2 + y_1 y_4 r_1^2 + y_1 y_3 r_1 r_2 + y_1 y_3 r_1^2 + y_1 y_2 r_1 r_2 + y_1 y_2 r_1^2 + y_1^2 r_2^2 + y_1^2 r_1 r_2 + y_1^4 t_4 r_1 + y_1^4 t_3 r_2 + y_1^4 t_3 r_1 + y_1^4 t_2 r_2 + y_1^4 t_2 r_1 + y_1^4 t_1 r_2 + y_1^4 t_1 r_1$

242. $q_3 q_5 = y_2 r_2 q_3 + y_2 r_1 q_4 + y_1 r_2 q_6 + y_1 r_2 q_4 + y_1 r_2 q_3 + y_1 r_2 q_2 + y_1 r_1 q_5 + y_1 r_1 q_4 + y_1 r_1 q_1 + y_3 y_4 r_2^2 + y_3 y_4 r_1 r_2 + y_2 y_3 r_2^2 + y_2 y_3 r_1 r_2 + y_2 y_3 r_1^2 + y_2^2 r_1^2 +$

$y_1y_4r_1r_2 + y_1y_3r_1r_2 + y_1y_2r_1r_2 + y_1y_2r_1^2 + y_1^2r_2^2 + y_1^2r_1r_2 + y_1^4t_4r_2 + y_1^4t_4r_1 + y_1^4t_2r_2 + y_1^4t_2r_1$

243. $q_3q_4 = y_2r_2q_3 + y_2r_1q_4 + y_1r_2q_6 + y_1r_2q_4 + y_1r_2q_3 + y_1r_2q_2 + y_1r_1q_6 + y_1r_1q_4 + y_1r_1q_3 + y_1r_1q_2 + y_2y_4r_2^2 + y_2y_4r_1^2 + y_2y_3r_2^2 + y_2y_3r_1^2 + y_2^2r_1r_2 + y_1y_4r_2^2 + y_1y_4r_1^2 + y_1y_3r_1r_2 + y_1y_3r_1^2 + y_1^2r_2^2 + y_1^2r_1^2 + y_1^4t_3r_2 + y_1^4t_3r_1$

244. $q_3^2 = y_2y_4r_2^2 + y_2y_4r_1^2 + y_2y_3r_2^2 + y_2y_3r_1^2 + y_2^2r_1^2 + y_1y_4r_2^2 + y_1y_4r_1^2 + y_1y_2r_2^2 + y_1y_2r_1^2 + y_1^2r_2^2 + y_1^2r_1^2$

245. $q_2q_6 = y_2r_2q_4 + y_2r_1q_4 + y_1r_2q_6 + y_1r_2q_5 + y_1r_2q_4 + y_1r_2q_2 + y_1r_1q_5 + y_1r_1q_4 + y_1r_1q_2 + y_2y_4r_2^2 + y_2y_4r_1^2 + y_2^2r_2^2 + y_2^2r_1r_2 + y_1y_4r_2^2 + y_1y_4r_1r_2 + y_1y_3r_2^2 + y_1^2r_1r_2 + y_1^2r_1^2 + y_1^4t_4r_1 + y_1^4t_3r_1 + y_1^4t_1r_2$

246. $q_2q_5 = y_2r_2q_4 + y_2r_1q_4 + y_1r_2q_5 + y_1r_2q_4 + y_2y_4r_2^2 + y_2y_4r_1^2 + y_2^2r_2^2 + y_2^2r_1r_2 + y_1y_3r_2^2 + y_1y_2r_1r_2 + y_1y_2r_1^2 + y_1^2r_2^2 + y_1^2r_1r_2 + y_1^4t_4r_2 + y_1^4t_4r_1 + y_1^4t_3r_2 + y_1^4t_3r_1$

247. $q_2q_4 = y_2r_2q_4 + y_2r_1q_4 + y_1r_2q_2 + y_1r_1q_6 + y_1r_1q_5 + y_1r_1q_4 + y_1r_1q_2 + y_2y_4r_2^2 + y_2y_4r_1^2 + y_2^2r_2^2 + y_2^2r_1r_2 + y_1y_4r_1r_2 + y_1y_4r_1^2 + y_1y_3r_1r_2 + y_1y_3r_1^2 + y_1y_2r_2^2 + y_1y_2r_1r_2 + y_1^4t_4r_2 + y_1^4t_3r_2 + y_1^4t_1r_1$

248. $q_2q_3 = y_2r_2q_4 + y_2r_1q_4 + y_1r_2q_6 + y_1r_2q_5 + y_1r_2q_4 + y_2y_4r_2^2 + y_2y_4r_1^2 + y_2^2r_2^2 + y_2^2r_1r_2 + y_1y_4r_2^2 + y_1y_4r_1r_2 + y_1y_3r_2^2 + y_1y_3r_1r_2 + y_1y_2r_1r_2 + y_1y_2r_1^2 + y_1^4t_4r_1 + y_1^4t_3r_1 + y_1^4t_1r_2$

249. $q_2^2 = 0$

250. $q_1q_6 = y_2r_2q_3 + y_2r_1q_4 + y_1r_2q_6 + y_1r_2q_2 + y_1r_2q_1 + y_1r_1q_6 + y_1r_1q_5 + y_1r_1q_4 + y_1r_1q_1 + y_3y_4r_2^2 + y_3y_4r_1r_2 + y_3y_4r_1^2 + y_2y_4r_2^2 + y_2y_4r_1r_2 + y_2y_3r_1^2 + y_2^2r_1r_2 + y_1y_4r_2^2 + y_1y_3r_1r_2 + y_1y_3r_1^2 + y_1y_2r_1r_2 + y_1y_2r_1^2 + y_1^2r_2^2 + y_1^2r_1^2 + y_1^4t_4r_2 + y_1^4t_3r_2 + y_1^4t_2r_2$

251. $q_1q_5 = y_2r_1q_4 + y_1r_2q_4 + y_1r_1q_6 + y_1r_1q_1 + y_3y_4r_2^2 + y_2y_4r_1^2 + y_2y_3r_2^2 + y_2y_3r_1^2 + y_2^2r_2^2 + y_2^2r_1r_2 + y_1y_4r_1r_2 + y_1y_4r_1^2 + y_1y_3r_2^2 + y_1y_3r_1r_2 + y_1y_3r_1^2 + y_1y_2r_2^2 + y_1^2r_2^2 + y_1^2r_1r_2 + y_1^4t_4r_2 + y_1^4t_3r_2 + y_1^4t_3r_1 + y_1^4t_2r_2 + y_1^4t_2r_1 + y_1^4t_1r_2$

252. $q_1q_4 = y_2r_2q_4 + y_2r_1q_3 + y_1r_2q_6 + y_1r_2q_5 + y_1r_2q_4 + y_1r_2q_3 + y_1r_2q_2 + y_1r_2q_1 + y_1r_1q_6 + y_1r_1q_4 + y_1r_1q_1 + y_3y_4r_2^2 + y_3y_4r_1r_2 + y_2y_4r_2^2 + y_2y_3r_2^2 + y_2y_3r_1r_2 + y_2y_3r_1^2 + y_2^2r_2^2 + y_2^2r_1^2 + y_1y_4r_1r_2 + y_1y_4r_1^2 + y_1y_3r_2^2 + y_1y_3r_1r_2 + y_1y_3r_1^2 + y_1y_2r_1r_2 + y_1y_2r_1^2 + y_1^2r_1r_2 + y_1^2r_1^2 + y_1^4t_4r_1 + y_1^4t_3r_2 + y_1^4t_3r_1 + y_1^4t_1r_2$

253. $q_1q_3 = y_2r_2q_4 + y_2r_2q_3 + y_2r_1q_3 + y_1r_2q_4 + y_1r_2q_3 + y_1r_2q_1 + y_1r_1q_5 + y_1r_1q_1 + y_3y_4r_2^2 + y_3y_4r_1r_2 + y_2y_4r_1^2 + y_2y_3r_1^2 + y_2^2r_2^2 + y_2^2r_1r_2 + y_2^2r_1^2 + y_1y_4r_1r_2 + y_1y_4r_1^2 + y_1y_3r_2^2 + y_1y_2r_1r_2 + y_1^2r_2^2 + y_1^2r_1^2 + y_1^4t_3r_2 + y_1^4t_2r_1 + y_1^4t_1r_2 + y_1^4t_1r_1$

254. $q_1q_2 = y_2r_1q_4 + y_1r_2q_6 + y_1r_1q_6 + y_1r_1q_5 + y_1r_1q_4 + y_2y_4r_1r_2 + y_2y_4r_1^2 + y_2^2r_1r_2 + y_1y_4r_2^2 + y_1y_4r_1^2 + y_1y_3r_1^2 + y_1^2r_2^2 + y_1^2r_1r_2 + y_1^4t_4r_2 + y_1^4t_3r_2 + y_1^4t_1r_2 + y_1^4t_1r_1$

255. $q_1^2 = y_2y_4r_1^2 + y_2y_3r_2^2 + y_2y_3r_1^2 + y_2^2r_2^2 + y_1y_4r_2^2 + y_1y_4r_1^2 + y_1y_3r_2^2 + y_1y_2r_1^2 + y_1^2r_1^2$

256. $q_6o_2 = y_1^2y_2r_2q_4 + y_1^2y_2r_1q_4 + y_1^3r_2q_5 + y_1^3r_2q_2 + y_1^3r_2q_1 + y_1^3r_1q_5 + y_1^3r_1q_4 + y_1^2y_3y_4r_1^2 + y_1^2y_2y_3r_1r_2 + y_1^2y_2^2r_2^2 + y_1^2y_2^2r_1^2 + y_1^3y_4r_2^2 + y_1^3y_4r_1^2 + y_1^3y_3r_1^2 + y_1^3y_2r_1^2 + y_1^4r_1r_2 + y_1^4r_1^2$

257. $q_6o_1 = y_1^2y_2r_2q_4 + y_1^2y_2r_2q_3 + y_1^3r_2q_3 + y_1^3r_1q_3 + y_1^2y_3y_4r_1^2 + y_1^2y_2y_4r_2^2 + y_1^2y_2y_4r_1r_2 + y_1^2y_2y_3r_1r_2 + y_1^2y_2^2r_2^2 + y_1^2y_2^2r_1^2 + y_1^3y_4r_2^2 + y_1^3y_4r_1r_2 + y_1^3y_4r_1^2 + y_1^4r_1r_2$

258. $q_5 o_2 = y_1^2 y_2 r_1 q_4 + y_1^3 r_2 q_4 + y_1^3 r_2 q_3 + y_1^3 r_2 q_2 + y_1^3 r_1 q_6 + y_1^3 r_1 q_1 + y_1^2 y_3 y_4 r_1^2 + y_1^2 y_2 y_4 r_2^2 + y_1^2 y_2 y_3 r_2^2 + y_1^2 y_2 y_3 r_1 r_2 + y_1^2 y_2 y_3 r_1^2 + y_1^2 y_2^2 r_1 r_2 + y_1^3 y_4 r_1 r_2 + y_1^3 y_4 r_1^2 + y_1^3 y_3 r_1 r_2 + y_1^3 y_2 r_2^2 + y_1^3 y_2 r_1 r_2 + y_1^4 r_2^2$

259. $q_5 o_1 = y_1^2 y_2 r_2 q_4 + y_1^2 y_2 r_2 q_3 + y_1^2 y_2 r_1 q_4 + y_1^2 y_2 r_1 q_3 + y_1^3 r_2 q_4 + y_1^3 r_2 q_3 + y_1^3 r_1 q_4 + y_1^3 r_1 q_3 + y_1^3 r_1 q_1 + y_1^2 y_3 y_4 r_1 r_2 + y_1^2 y_3 y_4 r_1^2 + y_1^2 y_2 y_4 r_2^2 + y_1^2 y_2 y_3 r_2^2 + y_1^2 y_2^2 r_2^2 + y_1^2 y_2^2 r_1^2 + y_1^3 y_4 r_1^2 + y_1^3 y_2 r_1 r_2 + y_1^4 r_2^2 + y_1^4 r_1 r_2 + y_1^4 r_1^2$

260. $q_4 o_2 = y_1^2 y_2 r_2 q_4 + y_1^3 r_2 q_5 + y_1^3 r_2 q_3 + y_1^3 r_2 q_2 + y_1^3 r_1 q_5 + y_1^3 r_1 q_4 + y_1^3 r_1 q_3 + y_1^3 r_1 q_2 + y_1^3 r_1 q_1 + y_1^2 y_2 y_4 r_1^2 + y_1^2 y_2^2 r_1^2 + y_1^3 y_4 r_1 r_2 + y_1^3 y_3 r_2^2 + y_1^3 y_3 r_1^2 + y_1^3 y_2 r_1^2 + y_1^4 r_1^2$

261. $q_4 o_1 = y_1^2 y_2 r_1 q_4 + y_1^3 r_2 q_4 + y_1^3 r_2 q_2 + y_1^3 r_2 q_1 + y_1^3 r_1 q_6 + y_1^3 r_1 q_5 + y_1^2 y_3 y_4 r_1 r_2 + y_1^2 y_2 y_4 r_2^2 + y_1^2 y_2 y_4 r_1 r_2 + y_1^2 y_2 y_4 r_1^2 + y_1^2 y_2^2 r_2^2 + y_1^2 y_2^2 r_1 r_2 + y_1^2 y_2^2 r_1^2 + y_1^3 y_4 r_2^2 + y_1^3 y_4 r_1^2 + y_1^3 y_3 r_2^2 + y_1^3 y_3 r_1^2 + y_1^3 y_2 r_1^2 + y_1^4 r_2^2 + y_1^4 r_1 r_2$

262. $q_3 o_2 = y_1^2 y_2 r_2 q_4 + y_1^2 y_2 r_1 q_4 + y_1^2 y_2 r_1 q_3 + y_1^3 r_2 q_6 + y_1^3 r_2 q_4 + y_1^3 r_2 q_2 + y_1^3 r_2 q_1 + y_1^3 r_1 q_6 + y_1^3 r_1 q_5 + y_1^3 r_1 q_3 + y_1^3 r_1 q_2 + y_1^3 r_1 q_1 + y_1^2 y_3 y_4 r_1 r_2 + y_1^2 y_3 y_4 r_1^2 + y_1^2 y_2 y_4 r_1^2 + y_1^2 y_2 y_3 r_1 r_2 + y_1^2 y_2 y_3 r_1^2 + y_1^2 y_2^2 r_2^2 + y_1^2 y_2^2 r_1 r_2 + y_1^2 y_2^2 r_1^2 + y_1^3 y_4 r_1^2 + y_1^3 y_3 r_2^2 + y_1^3 y_3 r_1 r_2 + y_1^3 y_2 r_1 r_2 + y_1^4 r_1 r_2$

263. $q_3 o_1 = y_1^2 y_2 r_2 q_4 + y_1^2 y_2 r_1 q_4 + y_1^3 r_2 q_6 + y_1^3 r_1 q_6 + y_1^3 r_1 q_3 + y_1^3 r_1 q_1 + y_1^2 y_3 y_4 r_1^2 + y_1^2 y_2 y_4 r_1^2 + y_1^2 y_2^2 r_2^2 + y_1^2 y_2^2 r_1 r_2 + y_1^3 y_4 r_2^2 + y_1^3 y_4 r_1^2 + y_1^3 y_3 r_1 r_2 + y_1^3 y_3 r_1^2 + y_1^3 y_2 r_2^2 + y_1^3 y_2 r_1 r_2 + y_1^4 r_2^2 + y_1^4 r_1^2$

264. $q_2 o_2 = y_1^2 y_2 r_1 q_4 + y_1^2 y_2 r_1 q_3 + y_1^3 r_2 q_6 + y_1^3 r_2 q_5 + y_1^3 r_2 q_4 + y_1^3 r_1 q_6 + y_1^3 r_1 q_4 + y_1^2 y_2^2 r_1 r_2 + y_1^2 y_2^2 r_1^2 + y_1^3 y_4 r_2^2 + y_1^3 y_4 r_1 r_2 + y_1^3 y_3 r_2^2 + y_1^3 y_3 r_1 r_2 + y_1^3 y_3 r_1^2 + y_1^3 y_2 r_2^2 + y_1^3 y_2 r_1 r_2 + y_1^3 y_2 r_1^2$

265. $q_2 o_1 = y_1^3 r_2 q_6 + y_1^3 r_2 q_5 + y_1^3 r_2 q_4 + y_1^3 r_1 q_6 + y_1^3 r_1 q_4 + y_1^3 r_1 q_3 + y_1^3 r_1 q_1 + y_1^3 y_4 r_2^2 + y_1^3 y_3 r_2^2 + y_1^3 y_3 r_1^2 + y_1^3 y_2 r_2^2 + y_1^3 y_2 r_1^2$

266. $q_1 o_2 = y_1^2 y_2 r_2 q_4 + y_1^2 y_2 r_2 q_3 + y_1^2 y_2 r_1 q_3 + y_1^3 r_2 q_5 + y_1^3 r_2 q_4 + y_1^3 r_2 q_3 + y_1^3 r_2 q_2 + y_1^3 r_2 q_1 + y_1^3 r_1 q_5 + y_1^3 r_1 q_1 + y_1^2 y_3 y_4 r_1 r_2 + y_1^2 y_3 y_4 r_1^2 + y_1^2 y_2 y_4 r_1^2 + y_1^2 y_2 y_3 r_2^2 + y_1^2 y_2 y_3 r_1^2 + y_1^2 y_2^2 r_2^2 + y_1^2 y_2^2 r_1 r_2 + y_1^3 y_4 r_2^2 + y_1^3 y_4 r_1^2 + y_1^3 y_2 r_1 r_2 + y_1^4 r_2^2 + y_1^4 r_1 r_2$

267. $q_1 o_1 = y_1^2 y_2 r_1 q_4 + y_1^2 y_2 r_1 q_3 + y_1^3 r_2 q_5 + y_1^3 r_2 q_3 + y_1^3 r_2 q_2 + y_1^3 r_1 q_5 + y_1^3 r_1 q_3 + y_1^3 r_1 q_1 + y_1^2 y_3 y_4 r_1 r_2 + y_1^2 y_2 y_4 r_1 r_2 + y_1^2 y_2 y_4 r_1^2 + y_1^2 y_2 y_3 r_2^2 + y_1^2 y_2^2 r_1 r_2 + y_1^3 y_4 r_1 r_2 + y_1^3 y_3 r_2^2 + y_1^3 y_3 r_1^2 + y_1^3 y_2 r_1 r_2 + y_1^4 r_1 r_2$

268. $o_2^2 = 0$

269. $o_1 o_2 = y_1^5 r_1 q_4$

270. $o_1^2 = 0$

A minimal Gröbner basis for the relations ideal consists of the above minimal relations, together with the following superfluous relations:

271. $y_1 y_2 y_3 y_4 = y_1^2 y_3 y_4 + y_1^2 y_2 y_3 + y_1^2 y_2^2 + y_1^4$

272. $y_1 y_2^2 y_4 = y_1^2 y_3 y_4 + y_1^2 y_2 y_4 + y_1^2 y_2 y_3 + y_1^2 y_2^2 + y_1^3 y_4 + y_1^3 y_3$

273. $y_2^2 v_3 = y_1 y_2 v_4 + y_1^2 v_2 + y_1^2 v_1$

274. $y_2^2 v_2 = y_1 y_2 v_3 + y_1^2 v_4 + y_1^2 v_2 + y_1^2 v_1$

275. $y_1^4 y_3 y_4 = 0$

276. $y_1^4 y_2 y_4 = 0$

277. $y_1^5 y_4 = 0$

278. $y_1^5 y_3 = 0$

279. $y_1^5 y_2 = 0$

280. $y_1^6 = 0$

281. $y_2^2 t_8 = y_1^2 t_8 + y_1^2 t_5 + y_1^2 y_2^2 v_4 + y_1^3 y_2 v_4 + y_1^4 v_4$

282. $y_2^2 t_7 = y_1^2 t_8 + y_1^2 t_3 + y_1^2 t_2 + y_1^3 y_2 v_3 + y_1^4 v_4 + y_1^4 v_2 + y_1^4 v_1$

283. $y_2^2 t_4 = y_1^2 t_8 + y_1^2 t_5 + y_1^2 t_4 + y_1^2 t_1 + y_1^4 v_3 + y_1^4 v_2 + y_1^4 v_1$

284. $y_2^2 t_3 = y_1^2 t_7 + y_1^2 t_6 + y_1^2 t_5 + y_1^2 t_4 + y_1^2 t_2 + y_1^2 y_2^2 v_4 + y_1^3 y_2 v_2 + y_1^4 v_2$

285. $y_1 y_2 t_8 = y_1^2 t_8 + y_1^2 t_5 + y_1^2 t_4 + y_1^2 t_3 + y_1^4 v_3$

286. $y_1 y_2 t_7 = y_1^2 t_8 + y_1^2 t_6 + y_1^2 t_5 + y_1^2 t_3 + y_1^2 t_1 + y_1^3 y_2 v_4 + y_1^3 y_2 v_3 + y_1^3 y_2 v_2 + y_1^4 v_4$

287. $y_1 y_2 t_4 = y_1^2 t_7 + y_1^2 t_5 + y_1^2 t_4 + y_1^2 t_2 + y_1^2 t_1 + y_1^2 y_2^2 v_4 + y_1^4 v_3 + y_1^4 v_2 + y_1^4 v_1$

288. $y_1 y_2 t_3 = y_1^2 t_6 + y_1^2 t_5 + y_1^2 t_4 + y_1^2 y_2^2 v_4 + y_1^3 y_2 v_3 + y_1^3 y_2 v_2 + y_1^4 v_4 + y_1^4 v_2$

289. $y_1^4 y_2 v_4 = 0$

290. $y_1^4 y_2 v_3 = 0$

291. $y_1^4 y_2 v_2 = 0$

292. $y_1^5 v_4 = 0$

293. $y_1^5 v_3 = 0$

294. $y_1^5 v_2 = 0$

295. $y_1^5 v_1 = 0$

296. $y_1^4 t_8 = y_1^4 t_2 + y_1^4 t_1$

297. $y_1^4 t_7 = y_1^4 t_2 + y_1^4 t_1$

298. $y_1^4 t_6 = y_1^4 t_4 + y_1^4 t_3$

299. $y_2^2 q_4 = y_1^2 q_4 + y_1^2 q_3 + y_2^2 y_4 r_2 + y_2^2 y_4 r_1 + y_1 y_2 y_3 r_2 + y_1 y_2^2 r_1 + y_1^2 y_4 r_2 + y_1^2 y_2 r_1 + y_1^3 r_1$

300. $y_2^2 q_3 = y_1^2 q_4 + y_2^2 y_4 r_2 + y_2^2 y_4 r_1 + y_1 y_2 y_3 r_1 + y_1 y_2^2 r_1 + y_1^2 y_4 r_2 + y_1^2 y_2 r_2 + y_1^2 y_2 r_1 + y_1^3 r_2$

301. $y_1^5 t_4 = 0$

302. $y_1^5 t_3 = 0$

303. $y_1^5 t_2 = 0$

304. $y_1^5 t_1 = 0$

305. $y_1^3 y_2 q_4 = y_1^4 q_4 + y_1^4 q_2 + y_1^4 q_1 + y_1^3 y_2 y_4 r_2 + y_1^3 y_2 y_4 r_1 + y_1^4 y_4 r_2 + y_1^4 y_3 r_2 + y_1^4 y_2 r_2 + y_1^4 y_2 r_1 + y_1^5 r_2 + y_1^5 r_1$

306. $y_1^3 y_2 q_3 = y_1^4 q_4 + y_1^4 q_2 + y_1^4 q_1 + y_1^3 y_2 y_4 r_2 + y_1^3 y_2 y_4 r_1 + y_1^4 y_4 r_2 + y_1^4 y_3 r_2 + y_1^4 y_2 r_2 + y_1^4 y_2 r_1 + y_1^5 r_2 + y_1^5 r_1$

307. $y_1^4 q_6 = y_1^4 q_4 + y_1^4 q_2 + y_1^4 q_1 + y_1^4 y_4 r_1 + y_1^4 y_3 r_2 + y_1^4 y_3 r_1 + y_1^4 y_2 r_2$

308. $y_1^4 q_5 = y_1^4 q_1 + y_1^4 y_4 r_2 + y_1^4 y_2 r_1$

309. $y_1^5 q_3 = 0$

310. $y_1^5 q_2 = 0$

311. $y_1^5 q_1 = 0$

Essential ideal There are 21 minimal generators:

1. $y_1^2 y_3 y_4$
2. $y_1^2 y_2 y_4$
3. $y_1^2 y_2 y_3$
4. $y_1^2 y_2^2$
5. $y_1^3 y_4$
6. $y_1^3 y_3$

7. $y_1^3 y_2$
8. y_1^4
9. $y_1 y_2 v_4 + y_1^2 v_4 + y_1^2 v_1$
10. $y_1 y_2 v_3 + y_1^2 v_4$
11. $y_1 y_2 v_2 + y_1^2 v_2$
12. $y_1^2 t_8$
13. $y_1^2 t_7$
14. $y_1^2 t_6$
15. $y_1^2 t_5$
16. $y_1^2 t_4$
17. $y_1^2 t_3$
18. $y_1^2 t_2$
19. $y_1^2 t_1$
20. $y_1 q_6 + y_1 q_4 + y_1 q_2 + y_1 y_3 r_1 + y_1 y_2 r_2$
21. $y_1 q_5 + y_1 q_1 + y_1 y_4 r_2 + y_1 y_2 r_1$

Nilradical There are 24 minimal generators:

y_4, y_3, y_2, y_1, v_4, v_3, v_2, v_1, t_8, t_7, t_6, t_5, t_4, t_3, t_2, t_1, q_6, q_5, q_4, q_3, q_2, q_1, o_2, o_1.

Completion information

For this computation the minimal resolution was constructed out to degree 22. In this degree the presentation of the cohomology ring reaches its final form, and Carlson's criterion detects that the presentation is complete.

This cohomology ring has dimension 2 and depth 2. A homogeneous system of parameters is

$h_1 = r_1$ in degree 8
$h_2 = r_2$ in degree 8

The first two terms h_1, h_2 constitute a regular sequence of maximal length. The first two terms h_1, h_2 constitute a complete Duflot-regular sequence. That is to say, their restrictions to the greatest central elementary abelian subgroup constitute a regular sequence of maximal length.

Essential ideal The essential ideal is free of rank 75 as a module over the polynomial algebra in h_1, h_2. The free generators are:

In degree 4:

1. $G_{01} = y_1^2 y_3 y_4$
2. $G_{02} = y_1^2 y_2 y_4$
3. $G_{03} = y_1^2 y_2 y_3$
4. $G_{04} = y_1^2 y_2^2$
5. $G_{05} = y_1^3 y_4$
6. $G_{06} = y_1^3 y_3$

7. $G_{07} = y_1^3 y_2$
8. $G_{08} = y_1^4$

In degree 5:

1. $G_{09} = y_1^3 y_3 y_4$
2. $G_{10} = y_1^3 y_2 y_4$
3. $G_{11} = y_1^4 y_4$
4. $G_{12} = y_1^4 y_3$
5. $G_{13} = y_1^4 y_2$
6. $G_{14} = y_1^5$

In degree 6:

1. $G_{15} = y_1 y_2 v_4 + y_1^2 v_4 + y_1^2 v_1$
2. $G_{16} = y_1 y_2 v_3 + y_1^2 v_4$
3. $G_{17} = y_1 y_2 v_2 + y_1^2 v_2$

In degree 7:

1. $G_{18} = y_1 y_2^2 v_4 + y_1^2 y_2 v_4 + y_1^2 y_2 v_1$
2. $G_{19} = y_1^2 y_2 v_4 + y_1^3 v_4 + y_1^3 v_1$
3. $G_{20} = y_1^2 y_2 v_3 + y_1^3 v_4$
4. $G_{21} = y_1^2 y_2 v_2 + y_1^3 v_2$
5. $G_{22} = y_1^3 v_4$
6. $G_{23} = y_1^3 v_3$
7. $G_{24} = y_1^3 v_2$
8. $G_{25} = y_1^3 v_1$

In degree 8:

1. $G_{26} = y_1^2 t_8$
2. $G_{27} = y_1^2 t_7$
3. $G_{28} = y_1^2 t_6$
4. $G_{29} = y_1^2 t_5$
5. $G_{30} = y_1^2 t_4$
6. $G_{31} = y_1^2 t_3$
7. $G_{32} = y_1^2 t_2$
8. $G_{33} = y_1^2 t_1$
9. $G_{34} = y_1^2 y_2^2 v_4$
10. $G_{35} = y_1^3 y_2 v_4$
11. $G_{36} = y_1^3 y_2 v_3$
12. $G_{37} = y_1^3 y_2 v_2$
13. $G_{38} = y_1^4 v_4$
14. $G_{39} = y_1^4 v_3$
15. $G_{40} = y_1^4 v_2$
16. $G_{41} = y_1^4 v_1$

In degree 9:

1. $G_{42} = y_1^3 t_8$
2. $G_{43} = y_1^3 t_7$
3. $G_{44} = y_1^3 t_6$
4. $G_{45} = y_1^3 t_4$
5. $G_{46} = y_1^3 t_3$
6. $G_{47} = y_1^3 t_2$
7. $G_{48} = y_1^3 t_1$

In degree 10:

1. $G_{49} = y_1 q_6 + y_1 q_4 + y_1 q_2 + y_1 y_3 r_1 + y_1 y_2 r_2$
2. $G_{50} = y_1 q_5 + y_1 q_1 + y_1 y_4 r_2 + y_1 y_2 r_1$
3. $G_{51} = y_1^4 t_4$
4. $G_{52} = y_1^4 t_3$
5. $G_{53} = y_1^4 t_2$
6. $G_{54} = y_1^4 t_1$

In degree 11:

1. $G_{55} = y_1 y_2 q_4 + y_1 y_2 y_4 r_2 + y_1 y_2 y_4 r_1 + y_1 y_2^2 r_2 + y_1^2 y_2 r_2 + y_1^2 y_2 r_1$
2. $G_{56} = y_1 y_2 q_3 + y_1 y_2 y_4 r_2 + y_1 y_2 y_4 r_1 + y_1 y_2^2 r_1$
3. $G_{57} = y_1^2 q_6 + y_1^2 q_4 + y_1^2 q_2 + y_1^2 y_3 r_1 + y_1^2 y_2 r_2$
4. $G_{58} = y_1^2 q_5 + y_1^2 q_1 + y_1^2 y_4 r_2 + y_1^2 y_2 r_1$
5. $G_{59} = y_1^2 q_4 + y_1^2 y_4 r_2 + y_1^2 y_4 r_1 + y_1^2 y_2 r_2 + y_1^3 r_2 + y_1^3 r_1$
6. $G_{60} = y_1^2 q_3 + y_1^2 y_4 r_2 + y_1^2 y_4 r_1 + y_1^2 y_2 r_1$
7. $G_{61} = y_1^2 q_2$
8. $G_{62} = y_1^2 q_1 + y_1^2 y_4 r_1 + y_1^2 y_3 r_2$

In degree 12:

1. $G_{63} = y_1^2 y_2 q_4 + y_1^2 y_2 y_4 r_2 + y_1^2 y_2 y_4 r_1 + y_1^2 y_2^2 r_2 + y_1^3 y_2 r_2 + y_1^3 y_2 r_1$
2. $G_{64} = y_1^2 y_2 q_3 + y_1^2 y_2 y_4 r_2 + y_1^2 y_2 y_4 r_1 + y_1^2 y_2^2 r_1$
3. $G_{65} = y_1^3 q_6 + y_1^3 q_4 + y_1^3 q_2 + y_1^3 y_3 r_1 + y_1^3 y_2 r_2$
4. $G_{66} = y_1^3 q_5 + y_1^3 q_1 + y_1^3 y_4 r_2 + y_1^3 y_2 r_1$
5. $G_{67} = y_1^3 q_4 + y_1^3 y_4 r_2 + y_1^3 y_4 r_1 + y_1^3 y_2 r_2 + y_1^4 r_2 + y_1^4 r_1$
6. $G_{68} = y_1^3 q_3 + y_1^3 y_4 r_2 + y_1^3 y_4 r_1 + y_1^3 y_2 r_1$
7. $G_{69} = y_1^3 q_2$
8. $G_{70} = y_1^3 q_1 + y_1^3 y_4 r_1 + y_1^3 y_3 r_2$

In degree 13:

1. $G_{71} = y_1^4 q_4$
2. $G_{72} = y_1^4 q_3$
3. $G_{73} = y_1^4 q_2$
4. $G_{74} = y_1^4 q_1$

In degree 14:

1. $G_{75} = y_1^5 q_4$

Every essential class squares to zero, but the essential ideal does not square to zero. Here are the relations which are not of the form $G_i.G_j = 0$:

1. $G_{17}G_{26} = G_{75}$
2. $G_{16}G_{30} = G_{75}$
3. $G_{16}G_{29} = G_{75}$
4. $G_{16}G_{28} = G_{75}$
5. $G_{16}G_{27} = G_{75}$
6. $G_{15}G_{31} = G_{75}$
7. $G_{05}G_{49} = G_{75}$
8. $G_{03}G_{49} = G_{75}$
9. $G_{01}G_{50} = G_{75}$

Poincaré series The Poincaré series is of the form $f(t)/(1 - t^8)^2$, with

$$f(t) = 1 + 4t + 8t^2 + 10t^3 + 12t^4 + 13t^5 + 16t^6 + 20t^7$$
$$+ 16t^8 + 13t^9 + 12t^{10} + 10t^{11} + 8t^{12} + 4t^{13} + t^{14}.$$

A.7 The Sylow 3-subgroup of A_9

G is the Sylow 3-subgroup of the alternating group A_9. It is Small Group number 7 of order 81.

G has rank 2, 3-rank 3 and exponent 9. Its centre has 3-rank 1.

The 4 maximal subgroups are: 3^{1+2}_+, 3^{1+2}_- (twice) and 3^4.

There are 2 conjugacy classes of maximal elementary abelian subgroups. They are of 3-rank 2, 3 respectively.

This cohomology ring was successfully calculated.

Ring structure

The cohomology ring has 16 generators:

1. y_1 in degree 1, a nilpotent element
2. y_2 in degree 1, a nilpotent element
3. x_1 in degree 2, a nilpotent element
4. x_2 in degree 2
5. x_3 in degree 2
6. w_1 in degree 3, a nilpotent element
7. w_2 in degree 3, a nilpotent element
8. w_3 in degree 3, a nilpotent element
9. v_1 in degree 4, a nilpotent element
10. v_2 in degree 4
11. u_1 in degree 5, a nilpotent element
12. u_2 in degree 5, a nilpotent element
13. t_1 in degree 6, a nilpotent element

14. t_2 in degree 6
15. t_3 in degree 6, a regular element
16. s in degree 7, a nilpotent element

There are 88 minimal relations:

1. $y_2^2 = 0$
2. $y_1 y_2 = 0$
3. $y_1^2 = 0$
4. $y_2 x_3 = 0$
5. $y_1 x_2 = 0$
6. $y_2 x_1 = 0$
7. $y_1 x_1 = 0$
8. $x_2 x_3 = 0$
9. $x_1 x_3 = 0$
10. $x_1 x_2 = -y_2 w_1$
11. $x_1^2 = 0$
12. $y_2 w_2 = 0$
13. $y_1 w_3 = 0$
14. $y_1 w_1 = 0$
15. $x_3 w_3 = 0$
16. $x_3 w_1 = 0$
17. $x_2 w_2 = -y_2 v_1$
18. $y_1 v_2 = 0$
19. $x_1 w_3 = y_2 v_1$
20. $x_1 w_2 = 0$
21. $x_1 w_1 = 0$
22. $y_1 v_1 = 0$
23. $x_3 v_2 = 0$
24. $x_3 v_1 = 0$
25. $x_2 v_1 = -w_1 w_3 - y_2 u_1$
26. $x_1 v_2 = -y_2 u_1$
27. $w_3^2 = 0$
28. $w_2 w_3 = 0$
29. $w_2^2 = 0$
30. $w_1 w_2 = 0$
31. $w_1^2 = 0$
32. $x_1 v_1 = 0$
33. $y_1 u_2 = 0$
34. $y_1 u_1 = 0$
35. $w_2 v_2 = -y_2 t_1$
36. $w_1 v_2 = x_2 u_1 - y_2 t_2$
37. $x_3 u_2 = 0$
38. $x_3 u_1 = 0$
39. $y_1 t_2 = 0$
40. $w_3 v_1 = y_2 t_1$

41. $w_2v_1 = 0$

42. $w_1v_1 = 0$

43. $x_1u_2 = y_2t_1$

44. $x_1u_1 = 0$

45. $y_1t_1 = 0$

46. $x_3t_2 = 0$

47. $v_1v_2 = -w_1u_2$

48. $x_3t_1 = 0$

49. $x_2t_1 = -w_1u_2 - y_2s$

50. $x_1t_2 = y_2w_3v_2 - y_2x_2u_2 - y_2x_2u_1$

51. $v_1^2 = 0$

52. $w_3u_1 = -w_1u_2 - y_2s - y_2w_3v_2 + y_2x_2u_2 + y_2x_2u_1$

53. $w_2u_2 = 0$

54. $w_2u_1 = 0$

55. $w_1u_1 = y_2w_3v_2 - y_2x_2u_2 - y_2x_2u_1$

56. $x_1t_1 = 0$

57. $y_1s = 0$

58. $v_2u_1 = w_3t_2 - x_2s - x_2w_3v_2 + x_2^2u_2 + x_2^2u_1 + y_2v_2^2 - y_2x_2t_2$

59. $w_2t_2 = -y_2w_3u_2 + y_2w_1u_2$

60. $w_1t_2 = -x_2w_3v_2 + x_2^2u_2 + x_2^2u_1 + y_2v_2^2 - y_2x_2t_2$

61. $x_3s = 0$

62. $v_1u_2 = 0$

63. $v_1u_1 = y_2w_3u_2 - y_2w_1u_2$

64. $w_3t_1 = y_2w_3u_2 - y_2w_1u_2$

65. $w_2t_1 = 0$

66. $w_1t_1 = -y_2w_3u_2 + y_2w_1u_2$

67. $x_1s = y_2w_3u_2 - y_2w_1u_2$

68. $v_2t_1 = -w_3s + x_2w_3u_2 - x_2w_1u_2 - y_2v_2u_2$

69. $v_1t_2 = x_2w_3u_2 - x_2w_1u_2 - y_2v_2u_2$

70. $u_2^2 = 0$

71. $u_1u_2 = w_3s - x_2w_3u_2 + x_2w_1u_2 + y_2v_2u_2 - y_2w_1t_3$

72. $u_1^2 = 0$

73. $v_1t_1 = 0$

74. $w_2s = 0$

75. $w_1s = -x_2w_3u_2 + x_2w_1u_2 + y_2v_2u_2$

76. $u_2t_2 = v_2s - x_2w_1t_3$

77. $u_1t_2 = -w_3v_2^2 + x_2v_2u_2 + x_2w_3t_2 - x_2^2s - x_2^2w_3v_2 + x_2^3u_2 + x_2^3u_1 + y_2v_2t_2 + y_2x_2v_2^2 - y_2x_2^2t_2 - y_2x_2^2t_3$

78. $u_2t_1 = y_2v_1t_3$

79. $u_1t_1 = 0$

80. $v_1s = 0$

81. $t_2^2 = -v_2^3 - x_2v_2t_2 - x_2^3t_3$

82. $t_1t_2 = w_3v_2u_2 - x_2w_3s + x_2^2w_3u_2 - x_2^2w_1u_2 - y_2v_2s - y_2x_2v_2u_2 + y_2x_2w_3t_3 + y_2x_2w_1t_3$

83. $t_1^2 = 0$
84. $u_2 s = -w_1 w_3 t_3 - y_2 u_1 t_3$
85. $u_1 s = -w_3 v_2 u_2 + x_2 w_3 s - x_2^2 w_3 u_2 + x_2^2 w_1 u_2 + y_2 v_2 s + y_2 x_2 v_2 u_2 - y_2 x_2 w_3 t_3 - y_2 x_2 w_1 t_3$
86. $t_2 s = -v_2^2 u_2 - x_2 v_2 s - x_2^2 w_3 t_3 - x_2^2 w_1 t_3 + y_2 x_2 v_2 t_3$
87. $t_1 s = 0$
88. $s^2 = 0$

A minimal Gröbner basis for the relations ideal consists of the above minimal relations, together with the following superfluous relation:

89. $w_1 w_3 u_2 = -y_2 w_3 s + y_2 x_2 w_3 u_2 - y_2 x_2 w_1 u_2$

Essential ideal Zero ideal.

Nilradical There are 11 minimal generators:
$y_2, y_1, x_1, w_3, w_2, w_1, v_1, u_2, u_1, t_1, s$.

Completion information

For this cohomology computation the minimal resolution was constructed out to degree 14. The presentation of the cohomology ring reaches its final form in degree 14. Carlson's criterion detects in degree 14 that the presentation is complete.

This cohomology ring has dimension 3 and depth 2. A homogeneous system of parameters is

$h_1 = t_3$ in degree 6
$h_2 = x_3 + x_2$ in degree 2
$h_3 = v_2$ in degree 4

The first two terms h_1, h_2 constitute a regular sequence of maximal length. The last term h_3 is annihilated by the class y_1.

The first term h_1 constitutes a complete Duflot-regular sequence. That is to say, its restriction to the greatest central elementary abelian subgroup constitutes a regular sequence of maximal length.

Essential ideal The essential ideal is the zero ideal.

Poincaré series $\dfrac{1 + 2t + 2t^2 + 3t^3 + 3t^4 + 2t^5 + 2t^6 + t^7}{(1 - t^2)(1 - t^4)(1 - t^6)}$

A.8 Small Group No. 16 of order 243

G is Small Group number 16 of order 243.
G has rank 2, 3-rank 3 and exponent 27. Its centre has 3-rank 1.

One of the four maximal subgroups is the abelian group $C_9 \times C_3 \times C_3$. The other three are isomorphic to Small Group number 6 of order 81.

There is one conjugacy class of maximal elementary abelian subgroups. Each such subgroup has 3-rank 3.

This cohomology ring was successfully calculated.

Ring structure

The cohomology ring has 17 generators:

1. y_1 in degree 1, a nilpotent element
2. y_2 in degree 1, a nilpotent element
3. x_1 in degree 2, a nilpotent element
4. x_2 in degree 2, a nilpotent element
5. x_3 in degree 2
6. w_1 in degree 3, a nilpotent element
7. w_2 in degree 3, a nilpotent element
8. v_1 in degree 4, a nilpotent element
9. v_2 in degree 4
10. u_1 in degree 5, a nilpotent element
11. u_2 in degree 5, a nilpotent element
12. t_1 in degree 6, a nilpotent element
13. t_2 in degree 6
14. t_3 in degree 6, a regular element
15. s_1 in degree 7, a nilpotent element
16. s_2 in degree 7, a nilpotent element
17. r in degree 8, a nilpotent element

There are 103 minimal relations:

1. $y_2^2 = 0$
2. $y_1 y_2 = 0$
3. $y_1^2 = 0$
4. $y_1 x_3 = 0$
5. $y_2 x_2 = -y_1 x_2$
6. $y_2 x_1 = y_1 x_2$
7. $y_1 x_1 = 0$
8. $x_2 x_3 = y_2 w_1$
9. $x_1 x_3 = 0$
10. $x_2^2 = 0$
11. $x_1 x_2 = 0$
12. $x_1^2 = 0$
13. $y_1 w_2 = 0$
14. $y_1 w_1 = 0$
15. $y_1 v_2 = 0$
16. $x_2 w_2 = y_2 v_1$

17. $x_2 w_1 = 0$
18. $x_1 w_2 = 0$
19. $x_1 w_1 = 0$
20. $y_1 v_1 = 0$
21. $x_3 v_1 = w_1 w_2 + y_2 x_3 w_2$
22. $x_2 v_2 = y_2 u_1 - y_2 x_3 w_2$
23. $x_1 v_2 = 0$
24. $w_2^2 = 0$
25. $w_1^2 = 0$
26. $x_2 v_1 = 0$
27. $x_1 v_1 = 0$
28. $y_1 u_2 = 0$
29. $y_1 u_1 = 0$
30. $w_2 v_2 = x_3 u_2 - x_3 u_1 + x_3^2 w_2 - x_3^2 w_1 + y_2 t_2 - y_2 t_1$
31. $w_1 v_2 = x_3 u_1 - x_3^2 w_2 - y_2 t_2 + y_2 t_1 - y_2 w_1 w_2$
32. $y_1 t_2 = 0$
33. $w_2 v_1 = 0$
34. $w_1 v_1 = -y_2 w_1 w_2$
35. $x_2 u_2 = y_2 t_1$
36. $x_2 u_1 = y_2 w_1 w_2$
37. $x_1 u_2 = 0$
38. $x_1 u_1 = 0$
39. $y_1 t_1 = 0$
40. $v_1 v_2 = w_1 u_2 + y_2 x_3 u_2 - y_2 x_3 u_1 + y_2 x_3^2 w_2 - y_2 x_3^2 w_1$
41. $x_3 t_1 = w_1 u_2 + y_2 s_2 + y_2 s_1$
42. $x_2 t_2 = y_2 s_1$
43. $x_1 t_2 = 0$
44. $v_1^2 = 0$
45. $w_2 u_2 = -w_1 u_2 - x_3 w_1 w_2$
46. $w_2 u_1 = -w_1 u_2 - y_2 s_2$
47. $w_1 u_1 = x_3 w_1 w_2 - y_2 s_1$
48. $x_2 t_1 = 0$
49. $x_1 t_1 = 0$
50. $y_1 s_2 = 0$
51. $y_1 s_1 = 0$
52. $w_2 t_2 = x_3 s_2 - y_2 v_2^2 + y_2 x_3 t_2 - y_2 x_3^2 v_2 + y_2 r + y_2 x_3 w_1 w_2$
53. $w_1 t_2 = x_3 s_1 + y_2 v_2^2 - y_2 x_3 t_2 + y_2 x_3^2 v_2 - y_2 r - y_2 w_1 u_2 - y_2 x_3 w_1 w_2$
54. $v_1 u_2 = -y_2 w_1 u_2 - y_2 x_3 w_1 w_2$
55. $v_1 u_1 = y_2 r - y_2 w_1 u_2$
56. $w_2 t_1 = y_2 r$
57. $w_1 t_1 = -y_2 r$
58. $x_2 s_2 = y_2 r$
59. $x_2 s_1 = 0$
60. $x_1 s_2 = 0$

61. $x_1 s_1 = 0$

62. $y_1 r = 0$

63. $v_2 t_1 = u_1 u_2 + x_3 w_1 u_2 + x_3^2 w_1 w_2 - y_2 x_3 s_1$

64. $v_1 t_2 = w_1 s_2 + y_2 v_2 u_1 + y_2 x_3 s_2 - y_2 x_3 s_1 - y_2 x_3^2 u_2 - y_2 x_3^2 u_1 + y_2 x_3^3 w_2 + y_2 x_3^3 w_1$

65. $x_3 r = w_1 s_2 - y_2 v_2 u_2 + y_2 v_2 u_1 - y_2 x_3 s_2 - y_2 x_3 s_1 - y_2 x_3^3 w_2 + y_2 x_3^3 w_1$

66. $u_2^2 = 0$

67. $u_1^2 = 0$

68. $v_1 t_1 = 0$

69. $w_2 s_2 = -y_2 v_2 u_2 + y_2 v_2 u_1 + y_2 x_3 s_2 + y_2 x_3^2 u_2 - y_2 x_3^3 w_1$

70. $w_2 s_1 = -w_1 s_2 + y_2 v_2 u_2 + y_2 v_2 u_1 - y_2 x_3 s_2 + y_2 x_3 s_1 + y_2 x_3^2 u_1 - y_2 x_3^3 w_2$

71. $w_1 s_1 = y_2 v_2 u_1 - y_2 x_3 s_1 - y_2 x_3^2 u_2 - y_2 x_3^2 u_1 + y_2 x_3^3 w_2 + y_2 x_3^3 w_1$

72. $x_2 r = 0$

73. $x_1 r = 0$

74. $u_2 t_2 = v_2 s_2 + v_2 s_1 + x_3^2 s_1 + y_2 x_3 v_2^2 - y_2 x_3^2 t_2 + y_2 x_3^3 v_2 - y_2 u_1 u_2 + y_2 x_3 w_1 u_2 + y_2 x_3^2 w_1 w_2$

75. $u_1 t_2 = v_2 s_1 + x_3^2 s_2 - y_2 v_2 t_2 + y_2 x_3 v_2^2 - y_2 x_3^2 t_2 + y_2 x_3^3 v_2 - y_2 x_3^2 t_3 - y_2 u_1 u_2 - y_2 w_1 s_2 - y_2 x_3^2 w_1 w_2$

76. $u_2 t_1 = -y_2 w_1 s_2$

77. $u_1 t_1 = y_2 w_1 s_2$

78. $v_1 s_2 = y_2 u_1 u_2 - y_2 w_1 s_2 - y_2 x_3 w_1 u_2 + y_2 x_3^2 w_1 w_2$

79. $v_1 s_1 = -y_2 u_1 u_2 + y_2 x_3 w_1 u_2 - y_2 x_3^2 w_1 w_2$

80. $w_2 r = -y_2 u_1 u_2$

81. $w_1 r = y_2 u_1 u_2 + y_2 w_1 s_2 - y_2 x_3^2 w_1 w_2$

82. $t_2^2 = -v_2^3 + x_3^2 v_2^2 + x_3^3 t_2 - x_3^4 v_2 - x_3^3 t_3 - u_1 s_2 - x_3^3 w_1 u_2 - x_3^3 w_1 w_2 - y_2 v_2 s_2 - y_2 x_3 v_2 u_2 - y_2 x_3 v_2 u_1 + y_2 x_3^3 u_2 + y_2 x_3^3 u_1 - y_2 x_3^4 w_2 - y_2 x_3^4 w_1 + y_2 x_3 w_1 t_3$

83. $t_1 t_2 = u_1 s_2 - y_2 v_2 s_1 + y_2 x_3 v_2 u_2 + y_2 x_3 v_2 u_1 - y_2 x_3^2 s_2 + y_2 x_3^2 s_1 + y_2 x_3^3 u_1 - y_2 x_3^4 w_2 - y_2 x_3 w_1 t_3$

84. $v_2 r = u_1 s_2 - y_2 v_2 s_2 - y_2 v_2 s_1 + y_2 x_3 v_2 u_2 + y_2 x_3 v_2 u_1 + y_2 x_3^2 s_2 + y_2 x_3 w_2 t_3$

85. $t_1^2 = 0$

86. $u_2 s_2 = u_1 s_2 + x_3 w_1 s_2 + y_2 v_2 s_2 - y_2 x_3 v_2 u_2 + y_2 x_3 v_2 u_1 + y_2 x_3^2 s_2 + y_2 x_3^3 u_2 - y_2 x_3^4 w_1 + y_2 x_3 w_2 t_3$

87. $u_2 s_1 = -u_1 s_2 - y_2 v_2 s_2 + y_2 x_3 v_2 u_2 - y_2 x_3^2 s_2 - y_2 x_3^2 s_1 + y_2 x_3^3 u_2 - y_2 x_3^3 u_1 + y_2 x_3^4 w_2 - y_2 x_3^4 w_1 - y_2 x_3 w_2 t_3$

88. $u_1 s_1 = -x_3 w_1 s_2 - y_2 v_2 s_1 + y_2 x_3 v_2 u_2 - y_2 x_3^2 s_2 - y_2 x_3^2 s_1 + y_2 x_3^3 u_2 - y_2 x_3^3 u_1 + y_2 x_3^4 w_2 - y_2 x_3^4 w_1 - y_2 x_3 w_1 t_3$

89. $v_1 r = 0$

90. $t_2 s_2 = -v_2^2 u_2 + v_2^2 u_1 + x_3^2 v_2 u_1 + x_3^3 s_2 + x_3^4 u_2 - x_3^4 u_1 + x_3^3 w_2 - x_3^5 w_1 + y_2 v_2^3 + y_2 x_3^2 v_2^2 + y_2 x_3^3 t_2 - x_3^2 w_2 t_3 + y_2 x_3^3 t_3 - y_2 x_3 u_1 u_2 - y_2 x_3 w_1 s_2 + y_2 x_3^3 w_1 w_2 - y_2 w_1 w_2 t_3$

91. $t_2 s_1 = -v_2^2 u_1 + x_3^2 v_2 u_2 + x_3^3 s_1 + x_3^4 u_1 - x_3^5 w_2 - y_2 v_2^3 - y_2 x_3 v_2 t_2 - y_2 x_3^2 v_2^2 - y_2 x_3^3 t_2 - x_3^2 w_1 t_3 - y_2 x_3^3 t_3 + y_2 u_1 s_2 + y_2 x_3^2 w_1 u_2 + y_2 w_1 w_2 t_3$

92. $t_1 s_2 = -y_2 u_1 s_2 - y_2 x_3 u_1 u_2 + y_2 x_3 w_1 s_2 + y_2 x_3^2 w_1 u_2 - y_2 x_3^3 w_1 w_2 - y_2 w_1 w_2 t_3$

93. $t_1 s_1 = y_2 u_1 s_2 + y_2 x_3 u_1 u_2 - y_2 x_3 w_1 s_2 - y_2 x_3^2 w_1 u_2 + y_2 x_3^3 w_1 w_2 + y_2 w_1 w_2 t_3$

94. $u_2 r = y_2 u_1 s_2 + y_2 x_3 u_1 u_2 + y_2 x_3 w_1 s_2 - y_2 x_3^2 w_1 u_2 - y_2 x_3^3 w_1 w_2$

95. $u_1 r = y_2 u_1 s_2 - y_2 x_3 u_1 u_2 + y_2 x_3 w_1 s_2 - y_2 w_1 w_2 t_3$

96. $t_2 r = -v_2 u_1 u_2 + x_3^2 w_1 s_2 + x_3^3 w_1 u_2 + y_2 v_2^2 u_2 - y_2 v_2^2 u_1 + y_2 x_3 v_2 s_2 - y_2 x_3 v_2 s_1 + y_2 x_3^3 s_2 + y_2 x_3^4 u_2 - y_2 x_3^5 w_1 - x_3 w_1 w_2 t_3 + y_2 x_3^2 w_2 t_3$

97. $s_2^2 = 0$

98. $s_1 s_2 = -v_2 u_1 u_2 + x_3^2 w_1 s_2 + x_3^3 w_1 u_2 - y_2 v_2^2 u_2 - y_2 x_3 v_2 s_2 + y_2 x_3^2 v_2 u_2 - y_2 x_3^2 v_2 u_1 + y_2 x_3^3 s_2 - y_2 x_3^4 u_1 + y_2 x_3^5 w_2 - x_3 w_1 w_2 t_3 - y_2 x_3^2 w_2 t_3 - y_2 x_3^2 w_1 t_3$

99. $s_1^2 = 0$

100. $t_1 r = 0$

101. $s_2 r = y_2 v_2 u_1 u_2 - y_2 x_3 u_1 s_2 - y_2 x_3^3 w_1 u_2 + y_2 x_3 w_1 w_2 t_3$

102. $s_1 r = y_2 v_2 u_1 u_2 + y_2 x_3 u_1 s_2 - y_2 x_3^3 w_1 u_2 + y_2 x_3 w_1 w_2 t_3$

103. $r^2 = 0$

These minimal relations constitute a Gröbner basis for the ideal of relations.

Essential ideal There is one minimal generator: $y_1 x_2$.

Nilradical There are 13 minimal generators:
$y_2, y_1, x_2, x_1, w_2, w_1, v_1, u_2, u_1, t_1, s_2, s_1, r$.

Completion information

For this computation the minimal resolution was constructed out to degree 16. In this degree the presentation of the cohomology ring reaches its final form, and Carlson's criterion detects that the presentation is complete.

This cohomology ring has dimension 3 and depth 1. A homogeneous system of parameters is

$h_1 = t_3$ in degree 6
$h_2 = x_3$ in degree 2
$h_3 = v_2$ in degree 4

The first term h_1 constitutes a regular sequence of maximal length. The two remaining terms h_2, h_3 are both annihilated by the class y_1.

The first term h_1 constitutes a complete Duflot-regular sequence. That is to say, its restriction to the greatest central elementary abelian subgroup constitutes a regular sequence of maximal length.

Essential ideal The essential ideal is free of rank 1 as a module over the polynomial algebra in h_1. The free generator is $G_1 = y_1 x_2$ in degree 3. The essential ideal squares to zero.

Poincaré series $\dfrac{1 + 2t + 2t^2 + 2t^3 + t^4 + t^5 + 2t^6 + 2t^7 + 2t^8 + t^9}{(1 - t^2)(1 - t^4)(1 - t^6)}$

Epilogue

There are many possibilties for further developing the methods presented here and performing new computations. Some are research projects in their own right, whereas others would only take a couple of days at most. However, the line had to be drawn somewhere for the current work.

We may draw the following conclusions:

- The new Gröbner bases for kG-modules pass the theoretical and practical parts of the test. They provide the most powerful method to date for constructing minimal resolutions over p-groups.
- Carlson's completeness criterion for the calculation of cohomology rings works just as well for p-groups with p odd as it does for 2-groups.
- There are p-groups such that some products of two essential classes are nonzero.

Areas for future work:

- Presumably it is straightforward to generalize the Gröbner basis methods for kG-modules to modules over finite dimensional basic algebras.
- Constructing further terms in the minimal resolution for the Sylow 2-subgroup of the Mathieu group M_{24} will be difficult, as an enormous amount of data has to be stored temporarily. Parallel computers offer one way of distributing the storage requirements. I can think of two possible strategies for parallelizing the construction of minimal resolutions, but their feasiblity has yet to be evaluated.
- As indicated in Remark 5.11, the calculation of product cocycles can be speeded up by calculating preimages under the first differential d_1 with a matrix rather than by reduction over a Gröbner basis. Once this has been implemented it must be possible to compute the cohomology ring of the Sylow 2-subgroup of the Conway group Co_3 out to degree 11.
- The storage management in the component of Diag that implements Gröbner bases for graded commutative algebras can almost certainly be improved. After this it might well be possible to finish off the computation of the cohomology ring of the Sylow 5-subgroup of the Conway group Co_1. It is the complexity of the relations ideal that has prevented a complete computation up to now.

– By Duflot's Theorem, the p-rank of the centre $Z(G)$ is a lower bound for the depth of the cohomology ring $H^*(G)$. For groups of small order, this lower bound is attained noticeably more often if $p \geq 5$ than for $p = 2, 3$. For groups of order p^3 there is one counterexample for each of $p = 2, 3$ but none for $p \geq 5$. For groups of order p^4 there are three counterexamples for each of $p = 2, 3$ but none for $p = 5$. However, the wreath product group $C_p \wr C_p$ of order p^{p+1} always has depth exceeding Duflot's lower bound [22]. It would be useful to have a group-theoretic description for depth.

– The current method for identifying a homogeneous system of parameters must be improved. Ideally one wants a system of low total dimension, which contains a regular sequence of maximal length. The paper [38] provides one way to implement these criteria using restriction information.

– We have not discussed the extent to which the choice of minimal generating set for the p-group G affects the performance of the construction of the minimal resolution. Experience shows however that the implications of this choice can be very great. One minimal generating set for the Sylow 2-subgroup of the Mathieu group M_{24} was successfully used to compute the minimal resolution out to the 9th term, whereas another did not even allow the construction of the 6th term.

– There are 2328 groups of order 128. With only a few exceptions it is probably possible to compute the cohomology ring of any one of these groups. But computing all 2328 cohomology rings would take a long time.

– Benson conjectures in [9] that the Castelnuovo–Mumford regularity of $H^*(G)$ is always zero, and proves this when the cohology ring has Cohen–Macaulay defect – that is, dimension minus depth – at most two.

　　To date, no defect three cohomology rings seem to have been calculated. The methods used in Diag for computing the homology of the Koszul complex are described in §6.4 and assume that the defect has been proved to be at most two. Carlson has shown [19] that the cohomology ring of each group of order 64 has defect at most two. By contrast, if G is small group number 52 of order 128, then $H^*(G)$ does have defect three. This is because G has p-rank four, its centre has rank one, and there is an essential class in degree four. So computations for groups of order 128 will provide the first serious test of the regularity conjecture.

– In the same paper [9], Benson also puts forward a new completion criterion, which it would be good to test.

– We could probably compute the cohomology rings of most of the 67 groups of order 243. But the extraspecial group 3^{1+4}_+ of exponent 3 will cause problems, as a particularly large number of terms of the resolution is needed.

– If G is a direct product group, then the Künneth theorem gives its cohomology. If G is not a direct product, then the essential ideal is free and finitely generated as a module over the polynomial algebra generated by a complete Duflot regular sequence [40]. This can allow one to conclude early on that there are no unknown relations between the known generators.

References

1. W. W. Adams and P. Loustaunau. *An introduction to Gröbner bases.* Graduate Studies in Math., vol. 3. American Mathematical Society, Providence, RI, 1994.
2. A. Adem, J. F. Carlson, D. B. Karagueuzian, and R. J. Milgram. The cohomology of the Sylow 2-subgroup of the Higman-Sims group. *J. Pure Appl. Algebra* **164** (2001), 275–305.
3. A. Adem and D. Karagueuzian. Essential cohomology of finite groups. *Comment. Math. Helv.* **72** (1997), 101–109.
4. A. Adem and R. J. Milgram. The mod 2 cohomology rings of rank 3 simple groups are Cohen-Macaulay. In F. Quinn, editor, *Prospects in topology (Princeton, NJ, 1994)*, Ann. of Math. Stud., vol. 138, pages 3–12. Princeton Univ. Press, Princeton, NJ, 1995.
5. D. J. Benson. *Representations and cohomology. I.* Cambridge Studies in Advanced Math., vol. 30. Cambridge Univ. Press, Cambridge, 1991. Second edition 1998.
6. D. J. Benson. *Representations and cohomology. II.* Cambridge Studies in Advanced Math., vol. 31. Cambridge Univ. Press, Cambridge, 1991. Second edition 1998.
7. D. J. Benson. *Polynomial invariants of finite groups.* London Math. Soc. Lecture Note Series, vol. 190. Cambridge Univ. Press, Cambridge, 1993.
8. D. J. Benson. Conway's group Co_3 and the Dickson invariants. *Manuscripta Math.* **85** (1994), 177–193.
9. D. J. Benson. Dickson invariants, regularity and computation in group cohomology. Preprint, Jan. 2003. arXiv: math.GR/0303187.
10. D. J. Benson and J. F. Carlson. The cohomology of extraspecial groups. *Bull. London Math. Soc.* **24** (1992), 209–235.
11. D. J. Benson and J. F. Carlson. Periodic modules with large period. *Quart. J. Math. Oxford* **43** (1992), 283–296.
12. G. M. Bergman. The diamond lemma for ring theory. *Adv. in Math.* **29** (1978), 178–218.
13. H. U. Besche, B. Eick, and E. A. O'Brien. The groups of order at most 2000. *Electron. Res. Announc. Amer. Math. Soc.* **7** (2001), 1–4.
14. H. U. Besche, B. Eick, and E. A. O'Brien. A millennium project: constructing small groups. *Internat. J. Algebra Comput.* **12** (2002), 623–644.
15. C. Broto and H.-W. Henn. Some remarks on central elementary abelian p-subgroups and cohomology of classifying spaces. *Quart. J. Math. Oxford* **44** (1993), 155–163.
16. J. F. Carlson. Depth and transfer maps in the cohomology of groups. *Math. Z.* **218** (1995), 461–468.

17. J. F. Carlson. Problems in the calculation of group cohomology. In Dräxler et al. [27], pages 107–120.

18. J. F. Carlson. Calculating group cohomology: Tests for completion. *J. Symbolic Comput.* **31** (2001), 229–242.

19. J. F. Carlson. *The Mod-2 Cohomology of 2-Groups.* Website, Department of Mathematics, Univ. of Georgia, Athens, GA, May 2001. (http://www.math.uga.edu/~lvalero/cohointro.html).

20. J. F. Carlson. The cohomology rings of the groups of order 64. Appendix to [23], in preparation.

21. J. F. Carlson, E. L. Green, and G. J. A. Schneider. Computing Ext algebras for finite groups. *J. Symbolic Comput.* **24** (1997), 317–325.

22. J. F. Carlson and H.-W. Henn. Depth and the cohomology of wreath products. *Manuscripta Math.* **87** (1995), 145–151.

23. J. F. Carlson and L. Townsley. *The Cohomology Rings of Finite Groups.* Kluwer, in preparation.

24. H. Cartan and S. Eilenberg. *Homological algebra.* Princeton Univ. Press, Princeton, N. J., 1956.

25. J. Clark. Mod-2 cohomology of the group $U_3(4)$. *Comm. Algebra* **22** (1994), 1419–1434.

26. J. H. Conway, R. T. Curtis, S. P. Norton, R. A. Parker, and R. A. Wilson. *Atlas of finite groups.* Oxford Univ. Press, Oxford, 1985.

27. P. Dräxler, G. O. Michler, and C. M. Ringel, editors. *Computational methods for representations of groups and algebras (Essen, 1997).* Birkhäuser Verlag, Basel, 1999.

28. J. Duflot. Depth and equivariant cohomology. *Comment. Math. Helv.* **56** (1981), 627–637.

29. W. G. Dwyer. Homology decompositions for classifying spaces of finite groups. *Topology* **36** (1997), 783–804.

30. W. G. Dwyer. Sharp homology decompositions for classifying spaces of finite groups. In A. Adem, J. Carlson, S. Priddy, and P. Webb, editors, *Group representations: cohomology, group actions and topology (Seattle, WA, 1996)*, pages 197–220. Amer. Math. Soc., Providence, RI, 1998.

31. D. Eisenbud. *Commutative Algebra with a View Toward Algebraic Geometry.* Graduate Texts in Math., vol. 150. Springer-Verlag, New York, 1995.

32. L. Evens. *The cohomology of groups.* Oxford Univ. Press, Oxford, 1991.

33. D. R. Farkas, C. D. Feustel, and E. L. Green. Synergy in the theories of Gröbner bases and path algebras. *Canad. J. Math.* **45** (1993), 727–739.

34. C. D. Feustel, E. L. Green, E. Kirkman, and J. Kuzmanovich. Constructing projective resolutions. *Comm. Algebra* **21** (1993), 1869–1887.

35. P. Fleischmann. Periodic simple modules for $SU_3(q^2)$ in the describing characteristic $p \neq 2$. *Math. Z.* **198** (1988), 555–568.

36. The GAP Group, *GAP – Groups, Algorithms, and Programming*, Version 4.3, 2002. (http://www.gap-system.org).

37. J. Grabmeier and L. A. Lambe. Computing resolutions over finite p-groups. In A. Betten, A. Kohnert, R. Laue, and A. Wassermann, editors, *Algebraic combinatorics and applications (Gößweinstein, 1999)*, pages 157–195. Springer, Berlin, 2001.

38. D. J. Green. On Carlson's depth conjecture in group cohomology. *Math. Z.* **244** (2003), 711–723.

39. D. J. Green. The essential ideal in group cohomology does not square to zero. Submitted, Feb. 2003. arXiv: math.GR/0302336.

40. D. J. Green. The essential ideal in group cohomology is often a free module. In preparation.

41. E. L. Green. Noncommutative Gröbner bases, and projective resolutions. In Dräxler et al. [27], pages 29–60.

42. E. L. Green. Multiplicative bases, Gröbner bases, and right Gröbner bases. *J. Symbolic Comput.* **29** (2000), 601–623.

43. E. L. Green, Ø. Solberg, and D. Zacharia. Minimal projective resolutions. *Trans. Amer. Math. Soc.* **353** (2001), 2915–2939.

44. J. Grodal. Higher limits via subgroup complexes. *Ann. of Math.* **155** (2002), 405–457.

45. M. Hall, Jr. and J. K. Senior. *The groups of order* 2^n ($n \leq 6$). The Macmillan Co., New York, 1964.

46. I. J. Leary. The mod-p cohomology rings of some p-groups. *Math. Proc. Cambridge Philos. Soc.* **112** (1992), 63–75.

47. T. Marx. The restriction map in cohomology of finite 2-groups. *J. Pure Appl. Algebra* **67** (1990), 33–37.

48. R. J. Milgram and M. Tezuka. The geometry and cohomology of M_{12}. II. *Bol. Soc. Mat. Mexicana* **1** (1995), 91–108.

49. P. A. Minh. Essential mod-p cohomology classes of p-groups: an upper bound for nilpotency degrees. *Bull. London Math. Soc.* **32** (2000), 285–291.

50. F. Mora. Groebner bases for noncommutative polynomial rings. In J. Calmet, editor, *Algebraic algorithms and error correcting codes (Grenoble, 1985)*, Lecture Notes in Computer Sci., vol. 229, pages 353–362. Springer, Berlin, 1986.

51. H. Mui. The mod p cohomology algebra of the extra-special group $E(p^3)$. Unpublished essay, 1982.

52. P. Nordbeck. On some basic applications of Gröbner bases in non-commutative polynomial rings. In B. Buchberger and F. Winkler, editors, *Gröbner bases and applications*, London Math. Soc. Lecture Note Series, vol. 251, pages 463–472. Cambridge Univ. Press, Cambridge, 1998.

53. T. Okuyama and H. Sasaki. Homogeneous systems of parameters in cohomology algebras of finite groups. *Arch. Math. (Basel)*, to appear.

54. J. Pakianathan and E. Yalçın. On nilpotent ideals in the cohomology ring of a finite group. *Topology* **42** (2003), 1155–1183.

55. D. Quillen. The spectrum of an equivariant cohomology ring. I, II. *Ann. of Math.* **94** (1971), 549–572 and 573–602.

56. M. Ringe. *The C MeatAxe*, Release 2.2.3. Lehrstuhl D für Math., RWTH Aachen, 1997. (http://www.math.rwth-aachen.de/LDFM/homes/MTX/).

57. D. J. Rusin. The cohomology of the groups of order 32. *Math. Comp.* **53** (1989), 359–385.

58. P. J. Webb. A local method in group cohomology. *Comment. Math. Helv.* **62** (1987), 135–167.

59. P. J. Webb. Constructing resolutions for p-groups. Conference talk (various versions), 2000–02. Most recent version currently available at http://www.math.umn.edu/~webb/MtHolyoke02.dvi.

60. R. A. Wilson, R. A. Parker, and J. N. Bray. *ATLAS of Finite Group Representations*. Website, School of Mathematics and Statistics, Univ. of Birmingham, Birmingham, 1996–2002. (http://www.mat.bham.ac.uk/atlas/).

Index

Vol. 1746: A. Degtyarev, I. Itenberg, V. Kharlamov, Real Enriques Surfaces. XVI, 259 pages. 2000.

Vol. 1747: L. W. Christensen, Gorenstein Dimensions. VIII, 204 pages. 2000.

Vol. 1748: M. Ruzicka, Electrorheological Fluids: Modeling and Mathematical Theory. XV, 176 pages. 2001.

Vol. 1749: M. Fuchs, G. Seregin, Variational Methods for Problems from Plasticity Theory and for Generalized Newtonian Fluids. VI, 269 pages. 2001.

Vol. 1750: B. Conrad, Grothendieck Duality and Base Change. X, 296 pages. 2001.

Vol. 1751: N. J. Cutland, Loeb Measures in Practice: Recent Advances. XI, 111 pages. 2001.

Vol. 1752: Y. V. Nesterenko, P. Philippon, Introduction to Algebraic Independence Theory. XIII, 256 pages. 2001.

Vol. 1753: A. I. Bobenko, U. Eitner, Painlevé Equations in the Differential Geometry of Surfaces. VI, 120 pages. 2001.

Vol. 1754: W. Bertram, The Geometry of Jordan and Lie Structures. XVI, 269 pages. 2001.

Vol. 1755: J. Azéma, M. Émery, M. Ledoux, M. Yor, Séminaire de Probabilités XXXV. VI, 427 pages. 2001.

Vol. 1756: P. E. Zhidkov, Korteweg de Vries and Nonlinear Schrödinger Equations: Qualitative Theory. VII, 147 pages. 2001.

Vol. 1757: R. R. Phelps, Lectures on Choquet's Theorem. VII, 124 pages. 2001.

Vol. 1758: N. Monod, Continuous Bounded Cohomology of Locally Compact Groups. X, 214 pages. 2001.

Vol. 1759: Y. Abe, K. Kopfermann, Toroidal Groups. VIII, 133 pages. 2001.

Vol. 1760: D. Filipović, Consistency Problems for Heath-Jarrow-Morton Interest Rate Models. VIII, 134 pages. 2001.

Vol. 1761: C. Adelmann, The Decomposition of Primes in Torsion Point Fields. VI, 142 pages. 2001.

Vol. 1762: S. Cerrai, Second Order PDE's in Finite and Infinite Dimension. IX, 330 pages. 2001.

Vol. 1763: J.-L. Loday, A. Frabetti, F. Chapoton, F. Goichot, Dialgebras and Related Operads. IV, 132 pages. 2001.

Vol. 1764: A. Cannas da Silva, Lectures on Symplectic Geometry. XII, 217 pages. 2001.

Vol. 1765: T. Kerler, V. V. Lyubashenko, Non-Semisimple Topological Quantum Field Theories for 3-Manifolds with Corners. VI, 379 pages. 2001.

Vol. 1766: H. Hennion, L. Hervé, Limit Theorems for Markov Chains and Stochastic Properties of Dynamical Systems by Quasi-Compactness. VIII, 145 pages. 2001.

Vol. 1767: J. Xiao, Holomorphic Q Classes. VIII, 112 pages. 2001.

Vol. 1768: M.J. Pflaum, Analytic and Geometric Study of Stratified Spaces. VIII, 230 pages. 2001.

Vol. 1769: M. Alberich-Carramiñana, Geometry of the Plane Cremona Maps. XVI, 257 pages. 2002.

Vol. 1770: H. Gluesing-Luerssen, Linear Delay-Differential Systems with Commensurate Delays: An Algebraic Approach. VIII, 176 pages. 2002.

Vol. 1771: M. Émery, M. Yor, Séminaire de Probabilités 1967-1980. A Selection in Martingale Theory. IX, 553 pages. 2002.

Vol. 1772: F. Burstall, D. Ferus, K. Leschke, F. Pedit, U. Pinkall, Conformal Geometry of Surfaces in S^4. VII, 89 pages. 2002.

Vol. 1773: Z. Arad, M. Muzychuk, Standard Integral Table Algebras Generated by a Non-real Element of Small Degree. X, 126 pages. 2002.

Vol. 1774: V. Runde, Lectures on Amenability. XIV, 296 pages. 2002.

Vol. 1775: W. H. Meeks, A. Ros, H. Rosenberg, The Global Theory of Minimal Surfaces in Flat Spaces. Martina Franca 1999. Editor: G. P. Pirola. X, 117 pages. 2002.

Vol. 1776: K. Behrend, C. Gomez, V. Tarasov, G. Tian, Quantum Comohology. Cetraro 1997. Editors: P. de Bartolomeis, B. Dubrovin, C. Reina. VIII, 319 pages. 2002.

Vol. 1777: E. García-Río, D. N. Kupeli, R. Vázquez-Lorenzo, Osserman Manifolds in Semi-Riemannian Geometry. XII, 166 pages. 2002.

Vol. 1778: H. Kiechle, Theory of K-Loops. X, 186 pages. 2002.

Vol. 1779: I. Chueshov, Monotone Random Systems. VIII, 234 pages. 2002.

Vol. 1780: J. H. Bruinier, Borcherds Products on O(2,1) and Chern Classes of Heegner Divisors. VIII, 152 pages. 2002.

Vol. 1781: E. Bolthausen, E. Perkins, A. van der Vaart, Lectures on Probability Theory and Statistics. Ecole d' Eté de Probabilités de Saint-Flour XXIX-1999. Editor: P. Bernard. VIII, 466 pages. 2002.

Vol. 1782: C.-H. Chu, A. T.-M. Lau, Harmonic Functions on Groups and Fourier Algebras. VII, 100 pages. 2002.

Vol. 1783: L. Grüne, Asymptotic Behavior of Dynamical and Control Systems under Perturbation and Discretization. IX, 231 pages. 2002.

Vol. 1784: L.H. Eliasson, S. B. Kuksin, S. Marmi, J.-C. Yoccoz, Dynamical Systems and Small Divisors. Cetraro, Italy 1998. Editors: S. Marmi, J.-C. Yoccoz. VIII, 199 pages. 2002.

Vol. 1785: J. Arias de Reyna, Pointwise Convergence of Fourier Series. XVIII, 175 pages. 2002.

Vol. 1786: S. D. Cutkosky, Monomialization of Morphisms from 3-Folds to Surfaces. V, 235 pages. 2002.

Vol. 1787: S. Caenepeel, G. Militaru, S. Zhu, Frobenius and Separable Functors for Generalized Module Categories and Nonlinear Equations. XIV, 354 pages. 2002.

Vol. 1788: A. Vasil'ev, Moduli of Families of Curves for Conformal and Quasiconformal Mappings.IX, 211 pages. 2002.

Vol. 1789: Y. Sommerhäuser, Yetter-Drinfel'd Hopf algebras over groups of prime order. V, 157 pages. 2002.

Vol. 1790: X. Zhan, Matrix Inequalities. VII, 116 pages. 2002.

Vol. 1791: M. Knebusch, D. Zhang, Manis Valuations and Prüfer Extensions I: A new Chapter in Commutative Algebra. VI, 267 pages. 2002.

Vol. 1792: D. D. Ang, R. Gorenflo, V. K. Le, D. D. Trong, Moment Theory and Some Inverse Problems in Potential Theory and Heat Conduction. VIII, 183 pages. 2002.

Vol. 1793: J. Cortés Monforte, Geometric, Control and Numerical Aspects of Nonholonomic Systems. XV, 219 pages. 2002.

Vol. 1794: N. Pytheas Fogg, Substitution in Dynamics, Arithmetics and Combinatorics. Editors: V. Berthé, S. Ferenczi, C. Mauduit, A. Siegel. XVII, 402 pages. 2002.

Vol. 1795: H. Li, Filtered-Graded Transfer in Using Noncommutative Gröbner Bases. IX, 197 pages. 2002.

Vol. 1796: J.M. Melenk, hp-Finite Element Methods for Singular Perturbations. XIV, 318 pages. 2002.

Recent Reprints and New Editions